C0-ATM-823

The Urbanization of Capital

ALSO BY THE AUTHOR

Explanation in Geography
Social Justice and the City
The Limits to Capital
Consciousness and the Urban Experience:
Studies in the History and Theory of Capitalist Urbanization

The Urbanization of Capital

Studies in the History and Theory of Capitalist Urbanization

DAVID HARVEY

The Johns Hopkins University Press
Baltimore, Maryland

Printed in Great Britain

First published, 1985, by the Johns Hopkins University Press,
701 West 40th Street, Baltimore, Maryland 21211

ACKNOWLEDGMENTS

Chapter 2, "The Geography of Capitalist Accumulation: Toward a Reconstruction of the Marxian Theory," has been rewritten from two articles on that theme which earlier appeared in *Antipode* 7, no. 2 (1975), and 13, no. 3 (1981), while Chapter 4, "Land Rent under Capitalism," has been revised out of an article in that same journal (14, no. 3 [1982]).

I want to thank Richard Peet not only for permission to use the material here but also for his sterling efforts in keeping that stimulating journal open as a publication outlet for radical thought in geography. Chapter 1, "The Urban Process under Capitalism: A Framework for Analysis," first appeared in *The International Journal of Urban and Regional Research* 2 (1978), published by Edward Arnold; Chapter 3, "Class-Monopoly Rent, Finance Capital, and the Urban Revolution," was published in *Regional Studies* 8 (1974), and has been slightly revised to put it in its historical context. I would like to thank the respective editors and publishers of these journals for permission to republish these essays here. Chapter 5, "Class Structure and the Theory of Residential Differentiation," is from *Bristol Essays in Geography*, edited by M. Chisholm and R. Peel and published by Heinemann Press; and Chapter 7, "On Planning the Ideology of Planning," was first published in *Planning in the 1980s: Challenge and Response*, edited by J. Burchall and published by the Center for Urban Affairs, Rutgers University Press, and has also been slightly revised to put it in its historical context. I thank the respective editors and publishers for permission to reprint these studies here. Figure 12 is taken with permission from B. J. L. Berry and E. Neils, "Location, Size, and Shape of Cities as Influenced by Environmental Factors: The Urban Environment Writ Large," in *The Quality of the Urban Environment*, ed. H. Perloff (Baltimore, 1969).

Library of Congress Cataloging in Publication Data

Harvey, David.
The urbanization of capital.
Bibliography: p.
Includes index.
1. Urban economics. 2. Capitalism – History.
3. Marxian school of sociology. 4. Marxian
economics. 5. Space in economics. I. Title.
HT321.H373 1985 307.7′6 85–9795
ISBN 0–8018–3144–X (alk. paper)

for john

It spoke with a hundred voices of that huge process of historic waste that the place in general keeps putting before you; but showing it in a light that drew out the harshness or the sadness, the pang . . . of the reiterated sacrifice to pecuniary profit.

– *Henry James*

Contents

Figures

Tables

Preface

It has been my ambition, ever since the writing of *Social Justice and the City*, to progress toward a more definitive Marxian interpretation of the history and theory of urbanization under capitalism than I there achieved. The studies collected together in these two companion books – *The Urbanization of Capital* and *Consciousness and the Urban Experience* – are markings down that path.

I turned to the Marxian categories in the early 1970s, and reaffirm my faith in them here, as the only ones suited to the active construction of rigorous, comprehensive, and scientific understandings of something as complex and rich as the historical geography of the urban process under capitalism. The manner of construction of this science also makes political sense. Rigorous science can never be neutral in human affairs; attempts to put oneself outside of history at best produce rigorous and well-meaning pseudosciences, of which positivism is surely the best example. But Marx also argues that the conscious struggle to create an alternative to capitalism – call it socialism or communism – has to be based on thorough material understandings of how capitalism works and how its workings naturally generate certain states of political and social consciousness. In order to change the world, he seems to say, we have to understand it. But that process cannot be understood one-sidedly. Who, Marx asks, is to educate the educators? Revolutionary understandings of the world cannot be had out of passive contemplation but arise through active struggle. Only through changing the world can we change ourselves. Our task, therefore, is not to understand the world but to change it. But that slogan cannot be read too one-sidedly either. Active reflection on our understandings, critique of bourgeois ideology, the struggle to make Marxian concepts both plain and hegemonic, and the evaluation of our own historical experience of struggle are as important activities as active engagement on the barricades. That is why Marx wrote *Capital*. And that is why I can write these words.

To seek an understanding of capitalist urbanization in Marxian terms is to resort, however, to a framework of understanding that is controversial, incomplete, and in some respects highly problematic. I sought to do something about the incompleteness in *The Limits to Capital.* I there tried to fill in all kinds of "empty boxes" in Marxian theory, such as the circulation of fixed capital and built environment formation; the appropriation of rent; the workings of money, finance, and credit; the production of monetary and financial crises; and the like. I needed to theorize such phenomena if I was ever to construct a comprehensive theory of urbanization. But, curiously, most reviewers passed by (mainly, I suspect, out of pure disciplinary prejudice) what I thought to be the most singular contribution of that work – the integration of the production of space and spatial configurations as an active element within the core of Marxian theorizing. That was the key theoretical innovation that allowed me to shift from thinking about history to historical geography and so to open the way to theorizing about the urban process as an active moment in the historical geography of class struggle and capital accumulation.

I readily confess, of course, that much of my fascination with the spatial dimension of human affairs comes out of my disciplinary background in geography. But if, as Giddens (1981) insists, time-space relations are "constitutive features of social systems," then the question of space is surely too important to be left exclusively to geographers. Social theorists of all stripes and persuasions should take it seriously. Yet there has been a strong and almost overwhelming predisposition to give time and history priority over space and geography. Marx, Weber, Durkheim, and Marshall all have that in common. We consequently lack, as Giddens goes on to observe, the conceptual apparatus "which would make space, and the control of space, integral to social theory." That lack is doubly disturbing. To begin with, the insertion of concepts of space and space relations, of place, locale, and milieu, into any of the various supposedly powerful but spaceless social and theoretic formulations has the awkward habit of paralyzing that theory's central propositions. Microeconomists working with perfect competition find only spatial monopoly and prices that fail to produce equilibrium; macroeconomists find as many economies as central banks and a great deal of guesswork affecting relations between them; sociologists find all sorts of "space-time edges" that disturb otherwise coherent processes of structuration; and Marxists, employing a vocabulary appropriate to universal class relations, find neighborhoods, communities, and nations that partition class struggle and capital accumulation into strange configurations of uneven geographical development. Whenever social theorists actively interrogate the meaning of geographical and spatial categories, either they are forced to so many ad hoc adjustments that their theory splinters into incoherency or they are forced to rework very basic propositions. Small wonder, then, that Saunders (1981, 278), in a recent attempt to save the supposed subdiscipline of urban sociology from such an ugly fate, offers the extraordinary proposition, for which no justification is or ever could be found, that "the problems of space . . . must be severed from concern with specific social processes."

Marxists cannot, unfortunately, claim any superior virtue on this score. One searches the major Marxist journals in vain for serious discussion of spatial concepts and geographical dimensionality. Marx himself is partly to blame for this state of affairs. He certainly gave priority to time over space and was not averse to dismissing the question of geographical variation as an "unnecessary complication." To be sure, he sometimes admitted – as I show at length in Chapter 2 – the importance of space and place, but this in no way compensates for theory that is powerful with respect to time but weak with respect to space. Historical materialism appeared to license the study of historical transformations while ignoring how capitalism produces its own geography. This left Lenin and the theorists of imperialism with a huge gap to fill. Unfortunately, they did so by ad hoc adjustments that permitted discussion of the development of capitalism in, say, Russia and India (as if such units made inherent sense) and provoked an alternative rhetoric of exploitation in which centers exploit peripheries, the First World subjugates the Third, and capitalist power blocs compete for domination of space (markets, labor supplies, raw materials, production capacity). But how can we reconcile the idea that people in one place exploit or struggle against those in another place with Marx's view of a capitalist dynamic powered by the exploitation of one class by another? Such ad hoc concessions to spatial questions as Lenin, Luxemburg, and the other theorists of imperialism introduced merely made the theoretical foundations of Marxism-Leninism ambiguous, sparking savage and often destructive disputes over the national question and the right to national self-determination, the significance of the urban-rural contradiction, the prospects for socialism in one country, the appropriate response to urban social movements, the importance of geographical decentralization, and the like. The ad hoc adjustments treated, unfortunately, of capitalism *in* space without considering how space is produced and how the processes of production of space integrate into the capitalist dynamic and its contradictions. Historical materialism has to be upgraded, I insist, to historical-geographical materialism. The historical geography of capitalism has to be the object of our theorizing.

That immediately poses the problem of the proper relation between historical geography (actually experienced) and theory. Much critical ado has been made about the supposed Marxist shortcomings in this regard, chiefly focusing on the enclosure of theory and evidence within such a coherent frame as to preclude "independent" verification. There are various levels of response to that criticism. Firstly, anyone who thinks that there is no problem in the way language of any sort captures experience and represents structures in the external world is flatulently and hypocritically preaching in the wind. The independence of data from theory is always relative. The choice, therefore, is between different modes of approach to a universal problem. There are, secondly, good reasons for preferring one kind of approach to another. The abstract theories of positivism, for example, must first be translated into working models (an exercise that necessarily encloses a representation of theory and data within the same frame) and then tested against data that are supposed to be samples of repetitive and independent events. Such a procedure is perfectly reasonable in relation to certain arenas of enquiry. But it is quite irrelevant to historical geography, which is a unique configuration of highly interdependent events in space and time. Measuring the growth of cities as if there were no trade, capital flow, migration, or cultural and political influence between them makes no sense whatsoever. For that reason, many historians, humanists, and historical geographers prefer to bury their theoretical and political orientations in the ambiguities of common language. Compared to the charming cacophony of that, positivism appears appealingly rigorous.

As a Marxist I am overtly rather than subliminally concerned with rigorous theory building in relation to unique configurations of historical-geographical processes. The theory building does not, however, take place in abstraction but entails a continuous dialogue between experience, action, concept formation, and dialectical theorizing. Since there is considerable and often heated debate in Marxist circles on such matters – including the celebrated polarization between the Althusserian structuralists and historians like E. P. Thompson – I should perhaps explain my own approach more overtly.

Theory construction has its own inner logic that deserves to be understood in its own terms. The tensions within a known conceptual apparatus, such as that which Marx bequeathed us, can be used to spin out new lines of argument so as to represent "as in a mirror" (to use Marx's phrase) the historical and geographical dynamic of a particular mode of production. We reach out dialectically (rather than inward deductively) to probe uncharted seas from a few seemingly secure islands of concepts. Different starting points yield different perspectives on the realities we seek to understand. What appears as a secure conceptual apparatus from one vantage point turns out to be partial and one-sided from another. But the construction of different theoretical windows helps us map the rich complexity of a mode of production with greater accuracy. Capitalism as viewed from the standpoint of production in the first volume of Marx's *Capital* looks very different from capitalism studied from the viewpoint of circulation in volume two. Bringing the two perspectives together (a project that Marx never completed) should give us a fuller picture of the structure of a mode of production and its inner contradictions. It is then possible to build upon an understanding of those contradictions and reach out to grasp successive resolutions and internalizations of those tensions within the realms of finance capital, the state apparatus, and the geography of uneven development (to cite some of the themes broached in *The Limits to Capital*).

The problem is then to bring these theoretical arguments into contact with experience in such a way as to build greater or lesser degrees of confidence into the veracity of the theoretical propositions advanced. I have never felt comfortable, however, with the idea that there is something called "experience" unmediated by imagination. We always approach the world with some well-honed conceptual apparatus, the capital equipment of our intellect, and interpret the world broadly in those terms. Yet there are, fortunately, moments, events, people, and experiences that impinge upon imagination in unexpected ways, that jolt and jar received ways of thinking and doing, that demand some extra imaginative or theoretical leap to give them meaning. Experience, furthermore, comes in many guises. Casual street interactions and observations, the reading of the local press and all the pamphlets that get thrust into one's hands at street corners, local political action and attempts at more national and international political collaboration all hang together in a muddle of often conflicting experiences. And then there is the literature – vast, rambling, diverse, sometimes purely rhetorical and polemical (and no less interesting for that) and at other times making claims to dry-as-dust science. The analyst has to sample all of that, wrestle with the ideas and information advanced, sometimes fight fiercely in intellectual combat with those who advance them. The literature is not purely academic either. Novels, plays, poems, songs, paintings, graffiti, photographs, architectural drawing and plans . . . all of these give clues, contain potential surprises. I confess my thinking on the urban process has been as much influenced by Dickens, Zola, Balzac, Gissing, Dreiser, Pynchon, and a host of others as it has been by urban historians.

But I find myself most deeply drawn to those works, of which I regard Engels's *Condition of the Working Class in England in 1844* as the most brilliant example, that function both as literature and social theory, as history and contemporary commentary. It was, I suspect, out of that admiration that I was drawn in the first instance to detailed studies of the Baltimore housing market (see Chap. 3) and later on to the transformation of Paris after 1848 and the production of the Commune of 1871 (reported on in depth in *Consciousness and the Urban Experience,* chaps. 3 and 4). Both offered rich mines from which to dig new insights to challenge theory. Yet those studies depended crucially for the manner of their formulation upon the prior existence of some kind of theoretical and conceptual frame upon which the historical reconstructions could be hung. Engels provided the frame for the Baltimore housing studies, and *The Limits to Capital* gave me a basis to investigate the transformation of Paris in Haussman's time.

The path between the historical and geographical grounding of experience and the rigors of theory construction is hard to negotiate. I would be foolish to deny that there is any constitutive danger of circularity and tautology here. To be sure I might, like anyone else (from positivists to humanists), see only what I want to see and merely reconstruct experience in theoretically given terms. But that danger exists in all arenas of research and is by no means confined to Marxism. So how, then, do I, as a Marxist, confront that problem? I do so by putting great emphasis upon individual and collective processes of reflection and speculation as mediating steps between theory and experience.

By speculation, I mean the interrogation of the conceptual apparatus through which experience is mediated, the adjustment of conceptual filters and the juggling of perspectives so as to create fresh windows and dimensions to the interpretation of experience. Marx's *Grundrisse* is for me a model of such a process. By reflection, I have in mind the evaluation of experience, a summing up that can point in new directions, pose new problems, and suggest fresh areas for historical and theoretical enquiry. I regard Marx's *Eighteenth Brumaire of Louis Bonaparte* as a model of that style of thinking. Reflection and speculation prepare the way for theory construction at the same time as they define an arena of open and fluid evaluation of theoretical conclusions. If verification has any formal meaning in the Marxist lexicon, it lies in the open and productive qualities of reflection and speculation in relation to political action. The eternal vigilance exercised through reflective and speculative thought is the only immediate safeguard that we have. In the long run, however, historical geography has its own way of negating theoretical perspectives that lack material relevance.

The studies assembled in these two companion works were written against a background of this overwhelming concern to bring theory and historico-geographical experience together in such a way as to illuminate both. Unfortunately, thematic considerations, coupled with the sheer volume of the materials, have dictated a division that reflects, more than it overcomes, the division between historical geography and theory. *The Urbanization of Capital* is biased more toward theorizing (though not totally so), whereas *Consciousness and the Urban Experience* is much more speculative and historical. While each book stands on its own, I would at least like to express the hope that they might be evaluated as a whole.

The thematic division has, however another origin. The studies on the urbanization of capital are primarily concerned with how labor, working under capitalist control, creates a "second nature" of built environments with particular kinds of spatial configurations. I am primarily concerned with how capitalism creates a physical landscape of roads, houses, factories, schools, shops, and so forth in its own image and what the contradictions are that arise out of such processes of producing space. This is an easier target for theorizing to the degree that studies of the circulation of capital in general can be broadened and disaggregated to encompass problems of fixed capital formation and circulation and the interventions of finance capital and appropriators of rent. But these processes of urbanization of capital are paralleled by the urbanization of social relations through, for example, the separation of workplace and living place, the reorganization of capitalist systems of production and control, the reorganization of consumption processes to meet capitalism's requirements, the fragmentation of social space in relation to labor market demands, and the like. The urbanization of capital is an objectification in the landscape of that intersection between the productive force of capital investment and the social relations required to reproduce an increasingly urbanized capitalism. But this implies that we should look also at the implications for political consciousness of such processes. The "urbanization of consciousness" has, I therefore submit, to be taken as a real social, cultural, and political phenomenon in its own right. But that topic is far harder to ground theoretically, at least given the theoretical apparatus available to us (in spite of the extraordinary efforts of thinkers like Gramsci and Lukacs). The studies on consciousness and the urban experience are therefore much more speculative and much more heavily reliant upon the detailed interrogation of historical-geographical experience.

But why choose the "urban" as a framework for analysis? It is, after all, but one of several spatial scales on which the production of space and of political consciousness might be examined – neighborhoods, regions, nation-states and power blocs being others. Indeed, there are many social theorists, including not a few Marxists, who reject the idea of urbanization as a "theoretically specific object of analysis." Examination of the urban process, it is said, can at best yield "real but relatively unimportant insights" into the workings of civil society (Saunders 1981). Even those like Giddens (1981, 147) who take the problem of space somewhat more seriously are prone to argue that "with the advent of capitalism, the city is no longer the dominant time-space container or 'crucible of power'; this role is assumed by the territorially bounded nation state." Only occasional mavericks like Jane Jacobs (1984) insist on privileging the urban as a unit of analysis before all else.

By focusing on urbanization I do not intend that it be considered a theoretically specific object of analysis separate from what capitalism is about. Capital, Marx insists, must be conceived of as a process and not reified as a thing. The study of urbanization is a study of that process as it unfolds through the production of physical and social landscapes and the production of consciousness. The study of urbanization is not the study of a legal, political entity or of a physical artifact. It is concerned with processes of capital circulation; the shifting flows of labor power, commodities, and money capital; the spatial organization of production and the transformation of space relations; movements of information and geopolitical conflicts between territorially-based class alliances; and so on. The fact that cities in the legal sense have lost political power and geopolitical influence or that distinctive urban economies now merge into megalopolitan concentrations is but a part of this urban process. And if that sounds vague and somewhat ambiguous compared to the usual reifications of urban studies, it is deliberately so. I prefer to keep the ambiguity open in order to concentrate on urbanization as a process rather than engaging in secure reifications that conceal rather than reveal the fluid processes at work. That way we can better integrate understandings of the urban process into broader conceptions of the dynamics of capitalism and understand how each is part and parcel of the other.

Personal and intellectual debts are always hard to tabulate. I had institutional support from the Department of Geography and Environmental Engineering in the Johns Hopkins University and I want to thank Professor Wolman for that. Carol Ehrlich of the Johns Hopkins Press was a delight to work with and Alison Richards of Basil Blackwell did a sterling job at the production stage. Many good friends in Baltimore helped me see, understand, and participate in things that might otherwise have passed me by. It is invidious to choose, perhaps, but let me extend personal thanks to Barbara Koeppel, Ric Pfeffer, Vicente Navarro, Cliff DuRand, and Chester Wickwire. And among those associated with Hopkins I want to mention Lata Chatterjee, Gene Mumy, Jörn Barnbrock, Amy Kaplan, and Erica Schoenberger and to give very special thanks to Dick Walker, Neil Smith, and Beatriz Nofal. I owe them all a tremendous debt.

The Urbanization of Capital

1

The Urban Process under Capitalism: A Framework for Analysis

My objective is to understand the urban process under capitalism. I confine myself to the capitalist forms of urbanization because I accept the idea that the "urban" has a specific meaning under the capitalist mode of production which cannot be carried over without a radical transformation of meaning (and of reality) into other social contexts.

Within the framework of capitalism, I hang my interpretation of the urban process on the twin themes of *accumulation* and *class struggle.* The two themes are integral to each other and have to be regarded as different sides of the same coin – different windows from which to view the totality of capitalist activity. The class character of capitalist society means the domination of labor by capital. Put more concretely, a class of capitalists is in command of the work process and organizes that process for the purposes of producing profit. The laborer, however, has command only over his or her labor power, which must be sold as a commodity on the market. The domination arises because the laborer must yield the capitalist a profit (surplus value) in return for a living wage. All of this is extremely simplistic, of course, and actual class relations (and relations between factions of classes) within an actual system of production (comprising production, services, necessary costs of circulation, distribution, exchange, etc.) are highly complex. The essential Marxian insight, however, is that profit arises out of the domination of labor by capital and that the capitalists as a class must, if they are to reproduce themselves, continuously expand the basis for profit. We thus arrive at a conception of a society founded on the principle of "accumulation for accumulation's sake, production for production's sake." The theory of accumulation which Marx constructs in *Capital* amounts to a careful enquiry into the dynamics of accumulation and an exploration of its contradictory character. This may sound rather economistic as a framework for analysis, but we have to recall that accumulation is the means whereby the capitalist class reproduces both itself and its domination over labor. Accumulation cannot, therefore, be isolated from class struggle.

I. THE CONTRADICTIONS OF CAPITALISM

We can spin a whole web of arguments concerning the urban process out of an analysis of the contradictions of capitalism. Let me set out the principal forms these contradictions take.

Consider, first, the contradiction that lies within the capitalist class itself. In the realm of exchange each capitalist operates in a world of individualism, freedom, and equality and can and must act spontaneously and creatively. Through competition, however, the inherent laws of capitalist production are asserted as "external coercive laws having power over every individual capitalist." A world of individuality and freedom on the surface conceals a world of conformity and coercion underneath. But the translation from individual action to behavior according to class norms is neither complete nor perfect – it never can be because the *process* of exchange under capitalist rules always presumes individuality, while the law of value always asserts itself in social terms. As a consequence, individual capitalists, each acting in his own immediate self-interest, can produce an aggregative result that is wholly antagonistic to the collective class interest. To take a rather dramatic example, competition may force each capitalist to so lengthen and intensify the work process that the capacity of the labor force to produce surplus value is seriously impaired. The collective effects of individual entrepreneurial activity can seriously endanger the social basis for future accumulation.

Consider, second, the implications of accumulation for the laborers. We know from the theory of surplus value that the exploitation of labor power is the source of capitalist profit. The capitalist form of accumulation therefore rests upon a certain violence that the capitalist class inflicts upon labor. Marx showed, however, that this appropriation could be worked out in such a way that it did not offend the rules of equality, individuality, and freedom as they must prevail in the realms of exchange. Laborers, like capitalists, "freely" trade the commodity they have for sale in the marketplace. But laborers are also in competition with each other for employment, while the work process is under the command of the capitalist. Under conditions of unbridled competition, the capitalists are forced willy-nilly into inflicting greater and greater violence upon those whom they employ. The individual laborer is powerless to resist this onslaught. The only solution is for the laborers to constitute themselves as a class and find collective means to resist the depredations of capital. The capitalist form of accumulation consequently calls into being overt and explicit class struggle between labor and capital. This contradiction between the classes explains much of the dynamic of capitalist history and is in many respects quite fundamental to understanding the accumulation process.

The two forms of contradiction are integral to each other. They express an underlying unity and are to be construed as different aspects of the same reality. Yet we can usefully separate them in certain respects. The internal contradiction within the capitalist class is rather different from the class confrontation between capital and labor, no matter how closely the two may be linked. In what follows I focus on the accumulation process in the absence of any overt response on the part of the working class to the violence that the capitalist class must necessarily inflict upon it. I then broaden the perspective and consider how the organization of the working class and its capacity to mount an overt class response affect the urban process under capitalism.

Various other forms of contradiction could enter in to supplement the analysis. For example, the capitalist production system often exists in an antagonistic relationship to non- or precapitalist sectors that may exist within (the domestic economy, peasant and artisan production sectors, etc.) or without it (precapitalist societies, socialist countries, etc.). We should also note the contradiction with "nature" which inevitably arises out of the relation between the dynamics of accumulation and the "natural" resource base as it is defined in capitalist terms. Such matters obviously have to be taken into account in any analysis of the history of urbanization under capitalism.

II. THE LAWS OF ACCUMULATION

I begin by sketching the structure of flows of capital within a system of production and realization of value. This I do with the aid of a series of diagrams which appear highly "functionalist" and perhaps unduly simple in structure, but which nevertheless help us to understand the basic logic of the accumulation process. We shall also see how problems arise because individual capitalists produce a result inconsistent with their class interest and consider some of the means whereby solutions to these problems might be found. In short, I attempt a summary of Marx's argument in *Capital* in the ridiculously short space of three or four pages.

The Primary Circuit of Capital

In volume one of *Capital,* Marx presents an analysis of the capitalist production process. The drive to create surplus value rests either on an increase in the length of the working day (absolute surplus value) or on the gains to be made from continuous revolutions in the "productive forces" through reorganizations of the work process which raise the productivity of labor power (relative surplus value). The capitalist captures relative surplus value from the organization of cooperation and division of labor within the work process or by the application of fixed capital (machinery). The motor for these continuous revolutions in the work process, for the rising productivity of labor, lies in capitalist competition as each capitalist seeks an excess profit by adopting a production technique superior to the social average.

The implications of all of this for labor are explored in a chapter entitled "The General Law of Capitalist Accumulation." Marx here examines alterations in the rate of exploitation and in the temporal rhythm of changes in the work process in relation to the supply conditions of labor power (in particular, the formation of an industrial reserve army), assuming all the while that a positive rate of accumulation must be sustained if the capitalist class is to reproduce itself. The analysis proceeds around a strictly circumscribed set of interactions, with all other problems assumed away or held constant. Figure 1 portrays the relations examined.

The second volume of *Capital* closes with a model of accumulation on an expanded scale. The problems of proportionality involved in the aggregative production of means of production and means of consumption are examined, with all other problems held constant (including technological change, investment in fixed capital, etc.). The objective here is to show the potential for crises of disproportionality within the production process. But Marx has now broadened the structure of relationships put under the microscope (fig. 2). Note, however, that in both cases Marx tacitly assumes that all commodities are produced and consumed within one time period. The structure of relations examined in figure 2 can be characterized as the *primary circuit of capital*.

Much of the analysis of the falling rate of profit and its countervailing tendencies in volume 3 similarly presupposes production and consumption within one time period, although there is some evidence that Marx intended to broaden the scope of this; however, he did not live to complete the work. But it is useful to consider the volume 3 analysis as a synthesis of the arguments presented in the first two volumes and as at the very least a cogent statement of the internal contradictions that exist within the primary circuit. Here we can clearly see the contradictions that arise out of the tendency for individual capitalists to act in a way that, when aggregated, runs counter to their own class interest. This contradiction produces a tendency toward *overaccumulation* – too much capital is produced in aggregate relative to the opportunities to employ that capital. This tendency is manifest in a variety of guises. We have:

1. Overproduction of commodities – a glut on the market.
2. Falling rates of profit (in pricing terms, to be distinguished from the falling rate of profit in value terms, which is a theoretical construct).

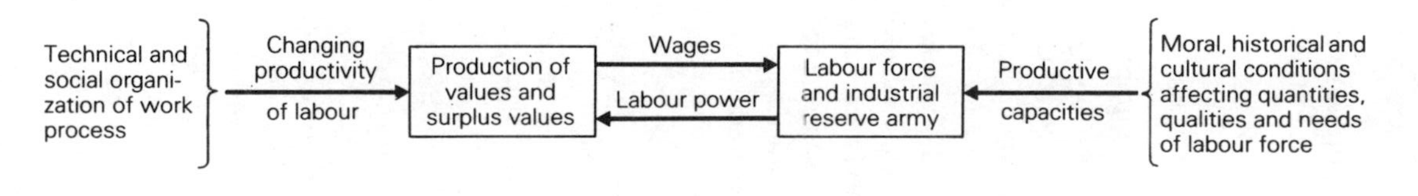

Fig. 1. *The relations considered in Marx's "general law of accumulation."*
(*Source:* Capital, *Vol. 1.*)

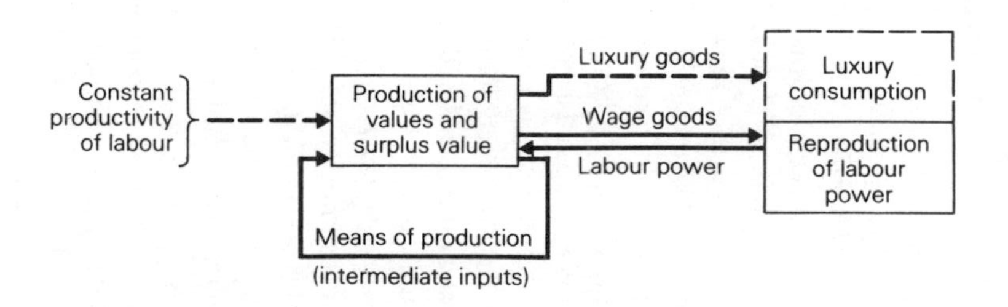

Fig. 2. *The relations considered in Marx's model of "reproduction on an expanded scale."*
(*Source:* Capital, *Vol. 2.*)

3. Surplus capital, which can be manifest either as idle productive capacity or as money capital lacking opportunities for profitable employment.
4. Surplus labor and/or a rising rate of exploitation of labor power.

One or a combination of these manifestations may be present simultaneously. We have here a preliminary framework for the analysis of capitalist crises (cf. Harvey 1982, chap. 7).

The Secondary Circuit of Capital

I now drop the tacit assumption of production and consumption within one time period and consider the problems posed by production and use of commodities requiring different working periods, circulation periods, and the like. This is an extraordinarily complex problem which Marx addresses to some degree in volume 2 of *Capital* and in the *Grundrisse.* Here I confine myself to some remarks regarding the formation of *fixed capital* and the *consumption fund.* Fixed capital, Marx argues, requires special analysis because of certain peculiarities that attach to its mode of production and realization. These peculiarities arise because fixed capital items can be produced in the normal course of capitalist commodity production, but they are used as aids to the production process rather than as direct raw material inputs. They are also used over a relatively long time period. We can also usefully distinguish between fixed capital enclosed within the production process and fixed capital that functions as a physical framework for production. The latter I call the *built environment for production.*

On the consumption side, we have a parallel structure. A *consumption fund* is formed out of commodities that function as aids rather than as direct inputs to consumption. Some items are directly enclosed within the consumption process (consumer durables such as stoves, washing machines, etc.), while others act as a physical framework for consumption (houses, sidewalks, etc.) – the latter I call the *built environment for consumption.*

We should note that some items in the built environment function jointly for both production and consumption – the transport network, for example – and that items can be transferred from one category to another by changes in use. Also, fixed capital in the built environment is immobile in space in the sense that the value incorporated in it cannot be moved without being destroyed. Investment in the built environment therefore entails the creation of a whole physical landscape for purposes of production, circulation, exchange, and consumption.

I call the capital flows into fixed asset and consumption fund formation the *secondary circuit of capital.* Consider, now, the manner in which such flows can occur. There must obviously be a "surplus" of both capital and labor in relation to current production and consumption needs in order to facilitate the movement of capital into the formation of long-term assets, particularly those constituting the built environment. The tendency toward overaccumulation produces such conditions within the primary circuit on a periodic basis. One feasible if *temporary* solution to this overaccumulation problem would therefore be to switch capital flows into the secondary circuit.

Individual capitalists will often find it difficult to bring about such a switch in flows for a variety of reasons. The barriers to individual switching of capital are particularly acute with respect to the built environment, where investments tend to be large-scale and long-lasting, often difficult to price in the ordinary way, and in many cases open to collective use by all individual capitalists. Indeed, individual capitalists left to themselves will tend to undersupply their own collective needs for production precisely because of such barriers. Individual capitalists tend to overaccumulate in the primary circuit and to underinvest in the secondary circuit; they have considerable difficulty in organizing a balanced flow of capital between the primary and secondary circuits.

A general condition for the flow of capital into the secondary circuit is, therefore, the existence of a functioning capital market and, perhaps, a state willing to finance and guarantee long-term, large-scale projects with respect to the creation of the built environment. At times of overaccumulation, a switch of flows from the primary to the secondary circuit can be accomplished only if the various manifestations of overaccumulation can be transformed into money capital that can move freely and unhindered into these forms of investment. This switch of resources cannot be accomplished without a money supply and credit system that creates "fictitious capital" *in advance* of actual production and consumption. This applies as much to the consumption fund (hence the importance of consumer credit, housing mortgages, municipal debt) as it does to fixed capital. Since the production of money and credit is a relatively autonomous process, we have to conceive of the financial and state institutions controlling the process as a kind of collective nerve center governing and *mediating* the relations between the primary and secondary circuits of capital. The nature and form of these financial and state institutions and the policies they adopt can play important roles in checking or enhancing flows of capital into the secondary circuit of capital or into certain specific aspects of it (such as transportation, housing, public facilities, and so on). An alteration in these mediating structures can therefore affect both the volume and the direction of capital flows by constricting movement down some channels and opening up new conduits elsewhere.

The Tertiary Circuit of Capital

In order to complete the picture of the circulation of capital in general, we have to conceive of a *tertiary circuit of capital* which comprises, first, investment in science and technology (the purpose of which is to harness science to production and thereby to contribute to the processes that continuously revolutionize the productive forces in society) and second, a wide range of social expenditures that relate primarily to the processes of reproduction of labor power. The latter can usefully be divided into investments directed toward the qualitative improvement of labor power from the standpoint of capital (investment in education and health by means of which the capacity of the laborers to engage in the work process will be enhanced) and investments in cooptation, integration, and repression of the labor force by ideological, military, and other means.

Individual capitalists find it hard to make such investments as individuals, no matter how desirable they may regard them. Once again, the capitalists are forced to some degree to constitute themselves as a class – usually through the agency of the state – and thereby to find ways to channel investment into research and development and into the quantitative and qualitative improvement of labor power. We should recognize that capitalists often *need* to make such investments in order to fashion an adequate social basis for further accumulation. But with regard to social expenditures, the investment flows are very strongly affected by the state of class struggle. The amount of investment in repression and in ideological control is directly related to the threat of organized working-class resistance to the depredations of capital. And the need to coopt labor arises only when the working class has accumulated sufficient power to require cooptation. Since the state can become a field of active class struggle, the mediations that are accomplished by no means fit exactly with the requirements of the capitalist class. The role of the state requires careful theoretical and historical elaboration in relation to the organization of capital flows into the tertiary circuit.

III. THE CIRCULATION OF CAPITAL AS A WHOLE AND ITS CONTRADICTIONS

Figure 3 portrays the overall structure of relations constituting the circulation of capital amongst the three circuits. The diagram looks very structuralist-functionalist because of the method of presentation. I can conceive of no other way to communicate clearly the various dimensions of capital flow. We now have to consider the contradictions embodied within these relations. I shall do so initially as if there were no overt class struggle between capital and labor. In this way we shall be able to see that the contradiction between the individual capitalist and capital in general is itself a source of major instability within the accumulation process.

We have already seen how the contradictions internal to the capitalist class generate a tendency toward overaccumulation within the primary circuit of capital. And I have argued that this tendency can be overcome, temporarily at

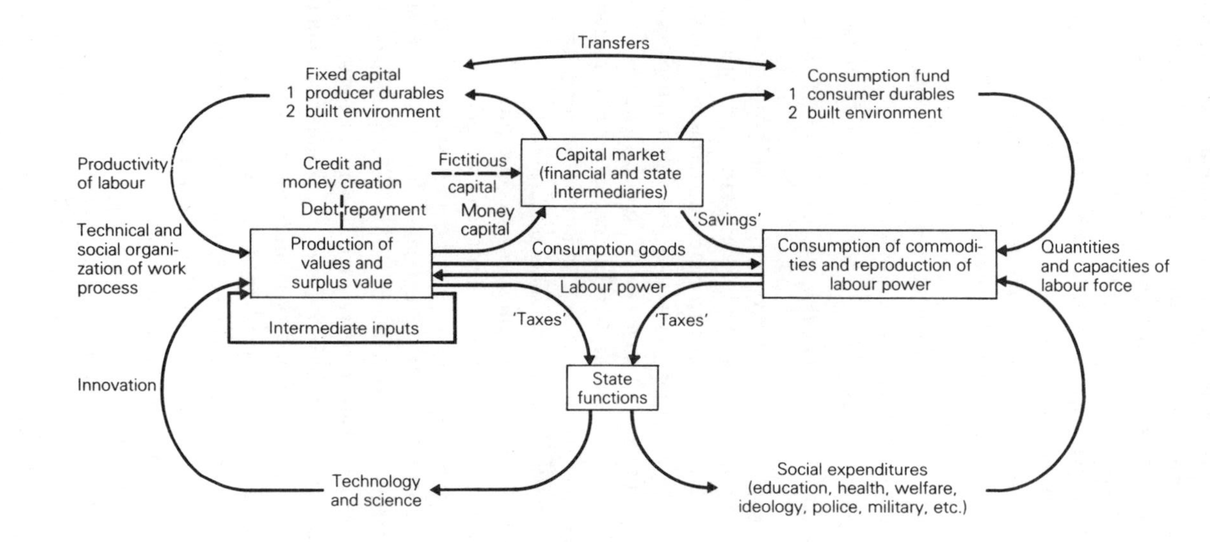

Fig. 3. The structure of relations between the primary, secondary, and tertiary circuits of capital

least, by switching capital into the secondary or tertiary circuits. Capital has, therefore, a variety of investment options open to it – fixed capital or consumption fund formation, investment in science and technology, investment in "human capital, as labor is usually called in bourgeois literature," or outright repression. At particular historical conjunctures capitalists may not be capable of taking up all of these options with equal vigor, depending upon the degree of their own organization, the institutions they have created, and the objective possibilities dictated by the state of production and the state of class struggle. I shall assume away such problems for the moment in order to concentrate on how the tendency toward overaccumulation, which I have identified so far only with respect to the primary circuit, manifests itself within the overall structure of circulation of capital. To do this I first need to specify the concept of productivity of investment.

On the Productivity of Investments in the Secondary and Tertiary Circuits

I choose the concept of "productivity" rather than "profitability" for a variety of reasons. First of all, the rate of profit as Marx treats of it in volume 3 of *Capital* is measured in value rather than pricing terms and takes no account of the distribution of the surplus value into its component parts of interest on money capital, profit on productive capital, rent on land, profit on merchants' capital, etc. The rate of profit is regarded as a social average earned by individual capitalists in all sectors, and it is assumed that competition effectively ensures its equalization. This is hardly a suitable conception for examining the flows between the three circuits of capital. To begin with, the formation of fixed capital in the built environment – particularly the collective means of production – cannot be understood without understanding the formation of a capital market and the distribution of part of the surplus in the form of interest. Second, many of the commodities produced in relation to the secondary and tertiary circuits cannot be priced in the ordinary way, while collective action by way of the state cannot be examined in terms of the normal criteria of profitability. Third, the rate of profit which holds is perfectly appropriate for understanding the behaviors of individual capitalists in competition but cannot be translated into a concept suitable for examining the behavior of capitalists as a class without some major assumptions (treating the total profit as equal to the total surplus value, for example).

The concept of productivity helps us to by-pass some of these problems if we specify it carefully enough. For the fact is that capitalists as a class – often through the agency of the state – do invest in the production of conditions that they hope will be favorable to accumulation, to their own reproduction as a class, and to their continuing domination over labor. This leads us immediately to a definition of a productive investment as one that directly or indirectly expands the basis for the production of surplus value. Plainly, investments in the secondary and tertiary circuits have the *potential* under certain conditions to do this. The problem – which besets the capitalists as much as it confuses us – is to identify the conditions and means that will allow this potential to be realized.

Investment in new machinery is the easiest case to consider. The new machinery is directly productive if it expands the basis for producing surplus value and unproductive if these benefits fail to materialize. Similarly, investment in science and technology may or may not produce new forms of scientific knowledge which can be applied to expand accumulation. But what of investment in roads, housing, health care and education, police forces and the military, and so on? If workers are being recalcitrant in the workplace, then judicious investment by the capitalist class in a police force to intimidate the workers and to break their collective power may indeed be productive indirectly of surplus value for the capitalists. If, however, the police are employed to protect the bourgeoisie in the conspicuous consumption of their revenues in callous disregard of the poverty and misery that surrounds them, then the police are not acting to facilitate accumulation. The distinction may be fine but it demonstrates the dilemma. How can the capitalist class identify, with reasonable precision, the opportunities for indirect and direct productive investment in the secondary and tertiary circuits of capital?

The main thrust of the modern commitment to planning (whether at the state or corporate level) rests on the idea that certain forms of investment in the secondary and tertiary circuits are potentially productive. The whole apparatus of cost-benefit analysis, of programming and budgeting, and of analysis of social benefits, as well as notions regarding investment in human capital, expresses this commitment and testifies to the complexity of the problem. And at the back of all of this is the difficulty of determining an appropriate basis for decision-making in the absence of clear and unequivocal profit signals. Yet the cost of bad investment decisions – investments that do not contribute directly or indirectly to accumulation of capital – must emerge somewhere. It must, as Marx would put it, come to the surface and thereby indicate the errors that lie beneath. We can begin to grapple with this question by considering the origins of crises within the capitalist mode of production.

On the Forms of Crisis under Capitalism

Crises are the real manifestation of the underlying contradictions within the capitalist process of accumulation. The argument that Marx puts forward throughout much of *Capital* is that there is always the potential within capitalism to achieve "balanced growth," but that this potential can never be realized because of the structure of the social relations prevailing in a capitalist society. This structure leads individual capitalists to produce results collectively that are antagonistic to their own class interest and leads them also to inflict an insupportable violence upon the working class which is bound to elicit its own response in the field of overt class struggle.

We have already seen how the capitalists tend to generate states of overaccumulation within the primary circuit of capital and have considered the various manifestations that result. As the pressure builds, either the accumulation process grinds to a halt or new investment opportunities are found as capital flows down various channels into the secondary and tertiary circuits. This movement may start as a trickle and become a flood as the potential for expanding the production of surplus value by such means becomes apparent. But the tendency toward overaccumulation is not eliminated. It is transformed, rather, into a pervasive tendency toward overinvestment in the secondary and tertiary circuits. This overinvestment is in relation solely to the needs of capital and has nothing to do with the real needs of people, which inevitably remain unfulfilled. Manifestations of crisis thus appear in both the secondary and tertiary circuits of capital.

As regards fixed capital and the consumption fund, the crisis takes the form of a crisis in the valuation of assets. Chronic overproduction results in the devaluation of fixed capital and consumption fund items – a process that affects the built environment as well as producer and consumer durables. We can likewise observe crisis formation at other points within the diagram of capital flows – crises in social expenditures (health, education, military repression, and the like), in consumption-fund formation (housing), and in technology and science. In each case the crisis occurs because the potential for productive investment within each of these spheres is exhausted. Further flows of capital do not expand the basis for the production of surplus value. We should also note that a crisis of any magnitude in any of these spheres is automatically registered as a crisis within the financial and state structures, while the latter, because of the relative autonomy that attaches to them, can be an independent source of crisis (we can thus speak of financial, credit, and monetary crises, the fiscal crises of the state, and so on).

Crises are the "irrational rationalizers" within the capitalist mode of production. They are indicators of imbalance and force a rationalization (which may be painful for certain sectors of the capitalist class as well as for labor) of the processes of production, exchange, distribution, and consumption. They may also force a rationalization of institutional structures (financial and state institutions in particular). From the standpoint of the total structure of relationships I have portrayed, we can distinguish different kinds of crises:

1. *Partial crises,* which affect a particular sector, geographical region, or set of mediating institutions. These can arise for any number of reasons but are

potentially capable of being resolved within that sector, region, or set of institutions. We can witness autonomously forming monetary crises, for example, which can be resolved by institutional reforms, crises in the formation of the built environment which can be resolved by reorganization of production for that sector, etc.

2. *Switching crises,* which involve a major reorganization and restructuring of capital flows and/or a major restructuring of mediating institutions in order to open up new channels for productive investments. It is useful to distinguish between two kinds of switching crises:

a. *Sectoral switching crises*, which entail switching the allocation of capital from one sphere (e.g. fixed capital formation) to another (e.g. education);
b. *Geographical switching crises,* which involve switching the flows of capital from one place to another. We note here that this form of crisis is particularly important in relation to investment in the built environment because the latter is immobile in space and requires interregional or international flows of money capital to facilitate its production.

3. *Global crises,* which affect, to a greater or lesser degree, all sectors, spheres, and regions within the capitalist production system. We will thus see devaluations of fixed capital and the consumption fund, a crisis in science and technology, a fiscal crisis in state expenditures, and a crisis in the productivity of labor, all manifest at the same time across all or most regions within the capitalist system. I note, in passing, that there have been only two global crises within the totality of the capitalist system – the first during the 1930s and its World War II, aftermath; the second, that which became most evident after 1973 but which had been steadily building throughout the 1960s.

A complete theory of capitalist crises should show how these various forms and manifestations relate in both space and time (see Harvey 1982). Such a task is beyond the scope of this chapter, but I can shed some light by returning to my fundamental theme – that of understanding the urban process under capitalism.

IV. ACCUMULATION AND THE URBAN PROCESS

The understanding I have to offer of the urban process under capitalism comes from seeing it in relation to the theory of accumulation. We must first establish the general points of contact between what seem, at first sight, two rather different ways of looking at the world.

Whatever else it may entail, the urban process implies the creation of a material physical infrastructure for production, circulation, exchange, and consumption. The first point of contact, then, is to consider the manner in which this built environment is produced and the way it serves as a resource system – a complex of use values – for the production of value and surplus value. We have, secondly, to consider the consumption aspect. Here we can usefully distinguish between the consumption of revenues by the bourgeoisie and the need to reproduce labor power. The former has a considerable impact upon the urban process, but I shall exclude it from the analysis because consideration of it would lead me into a lengthy discourse on the question of bourgeois culture and its complex significations without revealing very much directly about the specifically capitalist form of the urban process. Bourgeois consumption is, as it were, the icing on top of a cake that has as its prime ingredients capital and labor in dynamic relation to each other. The reproduction of labor power is essential and requires certain kinds of social expenditures and the creation of a consumption fund. The flows we have sketched, insofar as they portray capital movements into the built environment (for both production and consumption) and the laying out of social expenditures for the reproduction of labor power, provide us, then, with the structural links we need to understand the urban process under capitalism.

It may be objected, quite correctly, that these points of integration ignore the "rural-urban dialectic" and that the reduction of the urban process as we usually conceive of it to questions of built environment formation and reproduction of labor power is misleading if not downright erroneous. I would defend the reduction on a number of counts. First, as a practical matter, the mass of the capital flowing into the built environment and a large proportion of certain kinds of social expenditures are absorbed in areas that we usually classify as "urban." From this standpoint the reduction is a useful approximation. Second, I can discuss most of the questions that normally arise in urban research in terms of the categories of the built environment and social expenditures related to the reproduction of labor power with the added advantage that the links with the theory of accumulation can be clearly seen. Third, there are serious grounds for challenging the adequacy of the urban-rural dichotomy even when expressed as a dialetical unity, as a primary form of contradiction within the capitalist mode of production. In other words, and put quite bluntly, if the usual conception of the urban process appears to be violated by the reduction I am here proposing, then it is the usual conception of the urban process which is at fault.

The urban-rural dichotomy, for example, is regarded by Marx as an expression of the division of labor in society. In this, the division of labor is the fundamental concept and not the rural-urban dichotomy, which is just a particular form of its expression. Focusing on this dichotomy may be useful in seeking to understand social formations that arise in the transition to capitalism – such as those in which we find an urban industrial sector opposed to a rural peasant sector which is only formally subsumed within a system of commodity production and exchange. But in a purely capitalist mode of production – in which industrial and agricultural workers are all under the real domination of capital – this form of expression of the division of labor loses much of its particular significance. It disappears within a general concern for geographical specialization in the division of labor. And the other aspect of the urban process – the geographical concentration of labor power and use values for production and reproduction – also disappears quite naturally within an analysis of the rational spatial organization of physical and social infrastructures. In the context of advanced capitalist countries as well as in the analysis of the capitalist mode of production, the urban-rural distinction has lost its real economic basis, although it lingers, of course, within the realms of ideology with some important results. But to regard it as a fundamental conceptual tool for analysis is in fact to dwell upon a lost distinction that was in any case but a surface manifestation of the division of labor.

Overaccumulation and Long Cycles in Investment in the Built Environment

The acid test of any set of theoretical propositions comes when we seek to relate them to the experience of history and to the practices of politics. In the space of a chapter I cannot hope to demonstrate the relations between the theory of accumulation and its contradictions, on the one hand, and the urban process, on the other, in the kind of detail which would be convincing. I shall therefore confine myself to illustrating some of the more important themes that can be identified. I will focus first, exclusively on the processes governing investment in the built environment.

The system of production which capital established was founded on a physical separation between a place of work and a place of residence. The growth of the factory system, which created this separation, rested on the organization of cooperation, division of labor, and economies of scale in the work process as well as on the application of machinery. The system also promoted an increasing division of labor between enterprises and collective economies of scale through the agglomeration of activities in large urban centers. All of this meant the creation of a built environment to serve as a physical infrastructure for production, including an appropriate system for the transport of commodities. There are abundant opportunities for the productive employment of capital through the creation of a built environment for production. The same conclusion applies to investment in the built environment for consumption. The problem is, then, to discover how capital flows into the construction of this built environment and to establish the contradictions inherent in this process.

I should first say something about the concept of the built environment and consider some of its salient attributes. It is a complex composite commodity comprising innumerable different elements – roads, canals, docks and harbors, factories, warehouses, sewers, public offices, schools and hospitals, houses, offices, shops, etc. – each of which is produced under different conditions and according to quite different rules. The "built environment" is, then, a gross simplification, a concept that requires disaggregation as soon as we probe deeply into the processes of its production and use. Yet we also know that these components have to function as an ensemble in relation to the aggregative processes of production, exchange, and consumption. For purposes of exposition I can afford to remain at this level of generality. We also know that the built environment is long-lived, difficult to alter, spatially immobile, and often absorbent of large, lumpy investments. A proportion of it will be used in common by capitalists and consumers, and even those elements that can be privately appropriated (houses, factories, shops, etc.) are used in a context in which the externality effects of private uses are pervasive and often quite strong. All of these characteristics have implications for the investment process.

The analysis of fixed capital formation and the consumption fund in the context of accumulation suggests that investment in the built environment is likely to proceed according to a certain logic. Presume, for the moment, that the state does not take a leading role in promoting vast public works programs ahead of the demand for them. Individual capitalists, when left to their own devices, tend to underinvest in the built environment relative to their own individual and collective needs at the same time as they tend to overaccumulate. The theory then suggests that the overaccumulation can be siphoned off – via financial and state institutions and the creation of fictitious capital within the credit system – and put to work to make up the slack in investment in the built environment. This switch from the primary to the secondary circuit may occur in the course of a crisis or be accomplished relatively smoothly depending upon the efficiency of the mediating institutions. But the theory indicates that there is a limit to such a process and that at some point investments will become unproductive. At such a time the exchange value being put into the built environment has to be written down, diminished, or even totally lost. The fictitious capital contained within the credit system is seen to be just that, and financial and state institutions may find themselves in serious financial difficulty. The devaluation of capital in the built environment does not necessarily destroy the use value – the physical resource – constituted by the built environment. This physical resource can now be used as "devalued capital," and as such it functions as a free good that can help to reestablish the basis for renewed accumulation. From this we can see the logic of Marx's statement that periodic devaluations of fixed capital provide "one of the means immanent in capitalist production

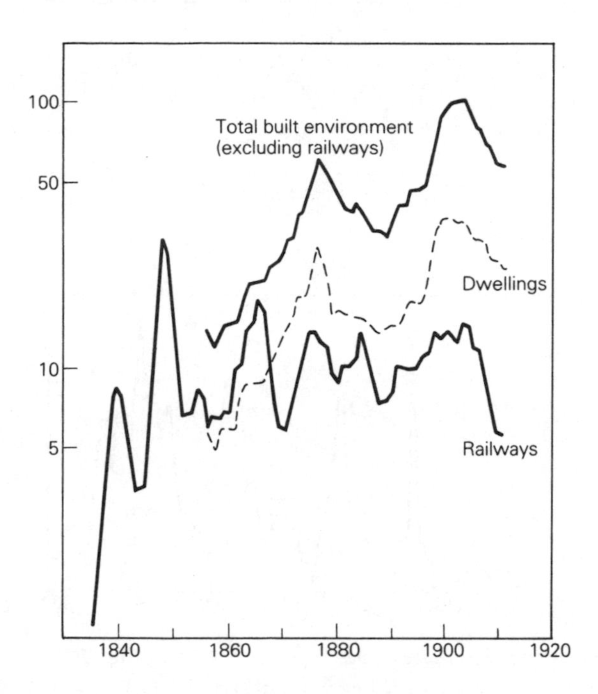

Fig. 4. Investment in selected components of the built environment in Britain, 1835–1914 (million £ at current prices)

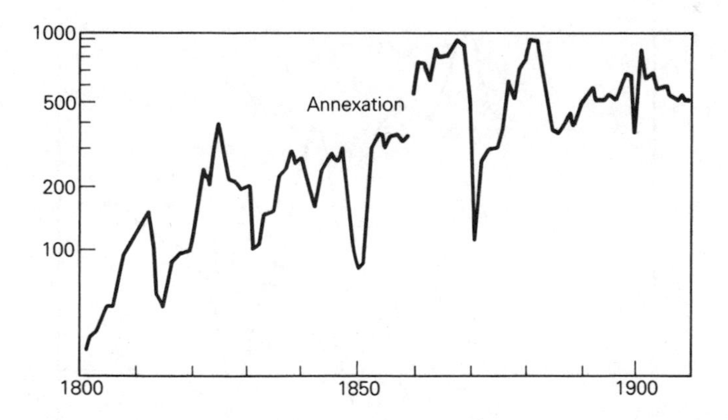

Fig. 5. Construction activity in Paris – entries of construction materials into the city, 1800–1910 (millions of cubic meters). (After Rougerie, 1968.)

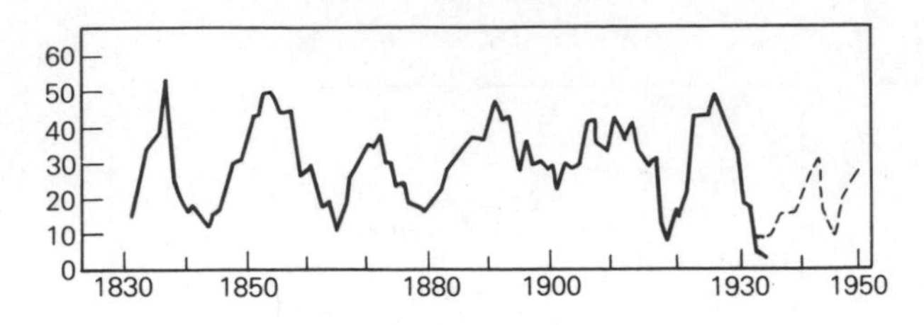

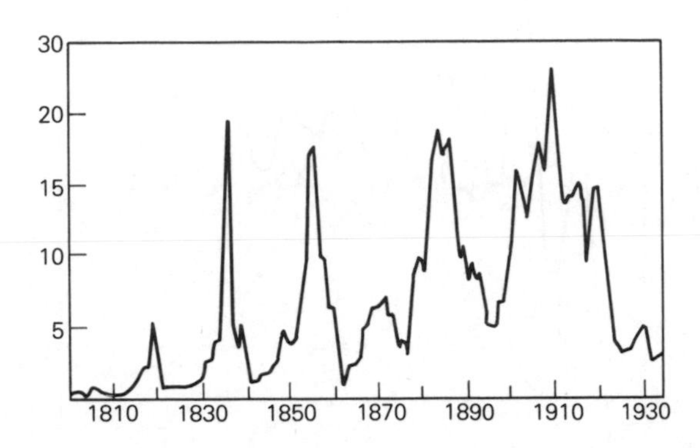

Fig. 6. Construction cycles in the United States, 1810–1950. Top: *Building activity per capita in the United States (1913 dollars per capita). (After B. Thomas, 1972.)* Bottom: *Sale of public lands in the United States (millions of acres of original land entries). (U.S. Department of Agriculture data.)*

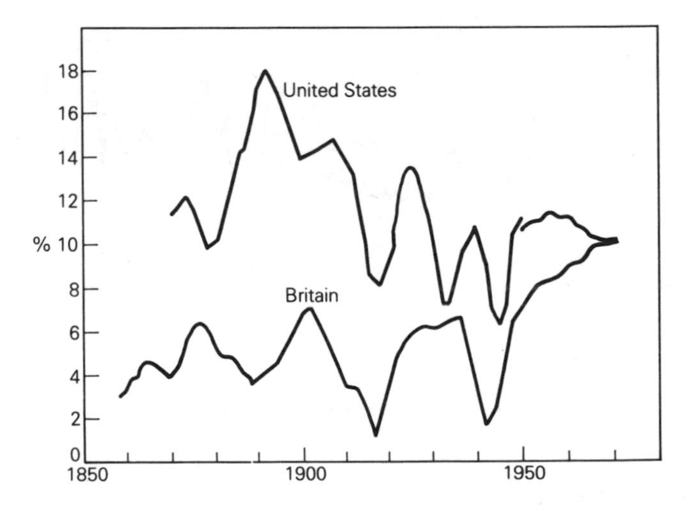

Fig. 7. Different rhythms of investment in the built environment in relation to GNP (U.S.A.) and GDP (Britain), 1860–1970 (five-year moving averages)

to check the fall of the rate of profit and hasten accumulation of capital-value through formation of new capital."

Since the impulses deriving from the tendency to overaccumulate and to underinvest are rhythmic rather than constant, we can construct a cyclical model of investment in the built environment. The rhythm is dictated in part by the rhythms of capital accumulation and in part by the physical and economic lifetime of the elements within the built environment – the latter means that change is bound to be relatively slow. The most useful thing we can do at this juncture is to point to the historical evidence for "long waves" in investment in the built environment. Somewhere in between the short-run movements of the business cycle – the "Juglar cycles" of approximately ten years' length – and the very long "Kondratieff cycles," we can identify movements of an intermediate length (sometimes called "Kuznets cycles"), which are strongly associated with waves of investment in the built environment. Gottlieb's recent investigation[1] of building cycles in thirty urban areas located in eight countries showed a periodicity clustering between fifteen and twenty-five years. While his methods and framework for analysis leave much to be desired, there is enough evidence accumulated by a variety of researchers to indicate that this is a reasonable sort of first-shot generalization. Figures 4, 5, and 6 illustrate the phenomenon. The historical evidence is at least consistent with my argument, taking into account, of course, the material characteristics of the built environment itself and in particular its long life, which means that instant throwaway cities are hardly feasible no matter how hard the folk in Los Angeles try.

The immobility in space also poses its own problematic with, again, its own appropriate mode of response. The historical evidence is, once more, illuminating. In the "Atlantic economy" of the nineteenth century, for example, the long waves in investment in the built environment moved inversely to each other in Britain and the United States (see figs. 7 and 8). The two movements were not independent of each other but were tied via migrations of capital and labor within the framework of the international economy at that time. The commercial crises of the nineteenth century switched British capital from home investment to overseas investment or vice versa. The capitalist "whole" managed, thereby, to achieve a roughly balanced growth through counterbalancing oscillations of the parts all encompassed within a global process of geographical expansion.[2] Uneven spatial development of the built environment was a crucial element in the achievement of relative global stability under the aegis of the *Pax Britannica*

[1] Gottlieb (1976) provides an extensive bibliography on the subject as well as his own statistical analysis. The question of long waves of various kinds has recently been brought back into the Marxist literature by Mandel (1975) and Day (1976).

[2] The main source of information is Brinley Thomas, *Migration and Economic Growth* (1972 edition), which has an extensive bibliography and massive compilations of data.

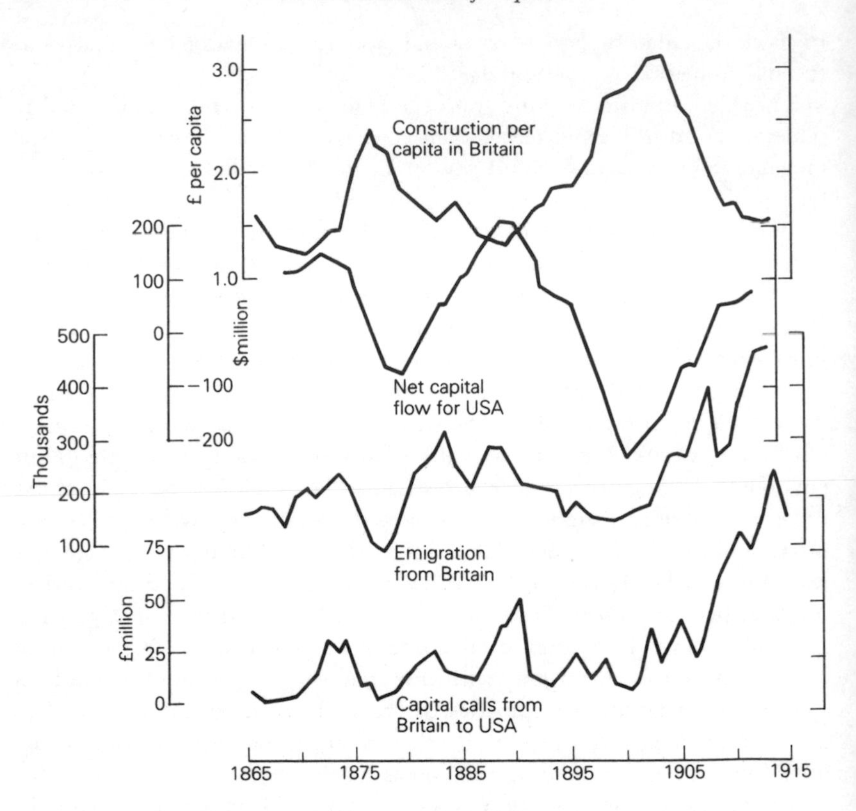

Fig. 8. Uneven development in the Atlantic economy, 1865–1914 – Britain and the United States. (After B. Thomas, 1972.)

of the nineteenth century. The crises of this period were either of the partial or of the switching variety, and we can spot both forms of the latter – geographical and sectoral – if we look carefully enough.

The global crises of the 1930s and the 1970s can in part be explained by the breakdown of the mechanisms for exploiting uneven development in this way. Investment in the built environment takes on a different meaning at such conjunctures. Each of the global crises of capitalism was in fact preceded by the massive movement of capital into long-term investment in the built environment as a kind of last-ditch hope for finding productive uses for rapidly overaccumulating capital. The extraordinary property boom in many advanced capitalist countries from 1969 to 1973, the collapse of which at the end of 1973 triggered (but did not cause) the onset of the current crisis, is a splendid example (see fig. 9).

While I am not attempting in any strict sense to verify the theory by

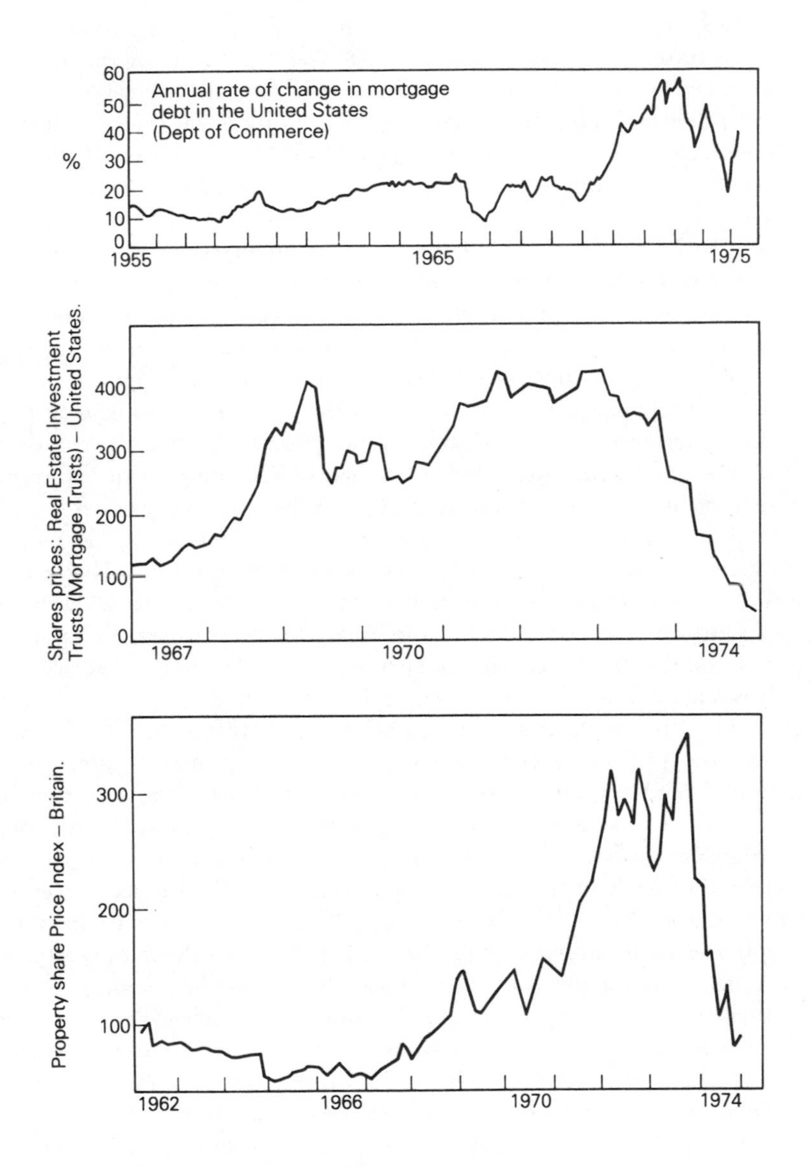

Fig. 9. Some indices of the property boom in Britain and the United States, 1955–1975.
Top: *Annual rate of change in mortgage debt in the United States. (Department of Commerce data.)* Middle: *Share prices of real estate investment trusts in the United States.*
(*Source:* Fortune Magazine.)
Bottom: *Property share price index in Britain.*
(*Source:* Investors Chronicle.)

appeal to the historical record, the latter most certainly is not incompatible with the broad outlines of the theory I have sketched. Bringing the theory to bear on the history is in fact an extraordinarily difficult task far beyond the scope of a short chapter. But rather than make no argument at all I shall try to illustrate how the connections can be made. I shall therefore look a little more closely at the two aspects of the theory which are crucial – overaccumulation and devaluation.

The flow of investment into the built environment depends upon the existence of surpluses of capital and labor and upon mechanisms for pooling the former and putting it to use. The history of this process is extremely interesting. The eighteenth century in Britain was characterized, for example, by a capital surplus, much of which went into the built environment because it had nowhere else to go. Investment in the built environment took place primarily for financial rather than use-value reasons – investors were looking for a steady and secure rate of return on their capital. Investment in property (much of it for conspicuous consumption by the bourgeoisie) and in turnpikes, canals, and rents (agricultural improvement) as well as in state obligations was about the only option open to rentiers. The various speculative crises that beset investment in the turnpikes and canals as well as urban property markets indicated very early that returns were by no means certain and that investments had to be productive if they were to succeed.[3]

It would be difficult to argue that during this period the surplus of capital arose out of the tendency to overaccumulate as I have specified it. The latter is, strictly speaking, a phenomenon that arises only in the context of the capitalist mode of production or in capitalist social formations that are relatively well developed. The long cycles of investment in the built environment predate the emergence of industrial capitalism and can be clearly identified throughout the transition from feudalism.[4] We can see, however, a strong relationship between these long cycles and fluctuations in the money supply and in the structure of capital markets. Perhaps the most spectacular example is that of the United States (see fig. 6) – when Andrew Jackson curbed land deals in paper currency and insisted on specie payment in 1836, the whole land development process came to a halt and the financial reverberations were felt everywhere, particularly by those investing in the built environment. The role of "fictitious" capital (see Harvey 1982, chaps. 9 and 10) and the credit and money supply system has always been fundamental

[3] The whole question of the capital surplus in the eighteenth century was first raised by Postan (1935) and subsequently elaborated on by Deane and Cole (1967). Recent studies on the financing of turnpikes and of canals in Britain by Albert (1972) and Ward (1974) provide some more detailed information.

[4] The best study is that by Parry Lewis (1965).

in relationship to the various waves of speculative investment in the built environment.

When, precisely, the tendency toward overaccumulation became the main agent producing surplus capital and when the long waves became explicitly tied to overaccumulation is a moot point. The evidence suggests that by the 1840s the connections had been strongly forged in Britain at least. By then, the functioning of the capital market was strongly bound to the rhythms imposed by the development of industrial capitalism. The "nerve center" that controls and mediates the relations between the primary and secondary circuits of capital increasingly functioned after 1830 or so according to a pure capitalist logic that affected both government and private activity. It is perhaps symptomatic that the fall of the July Monarchy in France in 1848 was directly related to the indebtedness of that regime incurred in the course of promoting a vast program of public works (many of which were not very productive). When the financial crisis, which had its origins in England and the extraordinary speculation in railroad construction, struck home in late 1846 and 1847, even the state debt of France could not withstand the shock.[5] For good reason, this crisis can perhaps be regarded as the first really solid and all-pervasive crisis in the capitalist world.

And what of the devaluation that inevitably results? If the devaluation is to function effectively, according to our theory, then it must leave behind a use value that can be used as the basis for further development. When many of the American states defaulted on their debts in the early 1840s, they failed to meet their obligations on the British capital market but kept the canals and other improvements that they had built. This was, in effect, expropriation without compensation – a prospect that the United States government treats with great moral indignation when some Third-World country threatens it today. The great railroad booms of the nineteenth century typically devalued capital while littering the landscape with physical assets that could usually be put to some use. When the urban mass transit systems went bankrupt at the turn of the century because of chronic overcapitalization, the mass transit systems were left behind as physical assets. Somebody had to pay for the devaluation, of course. There were the inevitable attempts to foist the costs onto the working class (often through municipal expenditures) or onto small investors. But big capital was not immune either, and the problems of the property companies in Britain or the real estate investment trusts in the United States in the years 1973–6 were exactly of this sort (although the involvement of pension funds and insurance companies affects individuals). The office space is still there, however, even though the building that houses it has been devalued and is now judged a nonearning asset. The history of

[5] See Girard (1952) and *The Urbanization of Consciousness*, chap. 3.

devaluations in the built environment is spectacular enough and fits, in general, with the theoretical argument.

The Contradictory Character of Investments in the Built Environment

I have so far treated the process of investment in the built environment as a mere reflection of the forces emanating from the primary circuit of capital. There are, however, a whole series of problems which arise because of the specific characteristics of the built environment itself. I shall consider these briefly.

Marx's extensive analysis of fixed capital in relation to accumulation reveals a central contradiction. On the one hand, fixed capital enhances the productivity of labor and thereby contributes to the accumulation of capital. But, on the other hand, it functions as a use value and requires the conversion of exchange values into a physical asset that has certain attributes. The exchange value locked up in this physical use value can be recouped only by keeping the use value fully employed over its lifetime, which for simplicity's sake I shall call its "amortization time." As a use value the fixed capital cannot easily be altered, and so it tends to freeze productivity at a certain level until the end of the amortization time. If new and more productive fixed capital comes into being before the old is amortized, then the exchange value still tied up in the old is devalued (Harvey 1982, chap. 8). Resistance to this devaluation checks the rise in productivity and, thus, restricts accumulation. The pursuit of new and more productive forms of fixed capital, however – dictated by the quest for relative surplus value – accelerates devaluations of the old.

We can identify exactly these same contradictory tendencies in relation to investment in the built environment, although they are even more exaggerated here because of the generally long amortization time involved, the fixity in space of the asset, and the composite nature of the commodity involved. I can demonstrate the argument most easily using the case of investment in transportation.

The cost, speed, and capacity of the transport system relate directly to accumulation because of the impacts these have on the turnover time of capital. Investment and innovation in transport are therefore potentially productive for capital in general. Under capitalism, consequently, we see a tendency to "drive beyond all spatial barriers" and to "annihilate space with time" (to use Marx's own expressions – see Chap. 2). This process is, of course, characterized typically by long cycles of the sort that we have already identified, uneven development in space, and periodic massive devaluations of capital.[6]

[6] See Isard (1942) for some interesting material.

I am here concerned, however, with the contradictions implicit in the process of transport development itself. Exchange values are committed to create "efficient" and "rational" configurations for spatial movement at a particular historical moment. There is, as it were, a certain striving toward spatial equilibrium, spatial harmony. In contrast, accumulation for accumulation's sake spawns continuous revolutions in transportation technology as well as a perpetual striving to overcome spatial barriers – all of which is disruptive of any existing spatial configuration.

We thus arrive at a paradox. In order to overcome spatial barriers and to annihilate space with time, spatial structures are created that themselves act as barriers to further accumulation. These spatial structures are expressed in the form of immobile transport facilities and ancillary facilities implanted in the landscape. We can in fact extend this conception to encompass the formation of the built environment as a whole. Capital represents itself in the form of a physical landscape created in its own image, created as use values to enhance the progressive accumulation of capital. The geographical landscape that results is the crowning glory of past capitalist development. But at the same time it expresses the power of dead labor over living labor, and as such it imprisons and inhibits the accumulation process within a set of specific physical constraints. And these can be removed only slowly unless there is a substantial devaluation of the exchange value locked up in the creation of these physical assets.

Capitalist development has therefore to negotiate a knife-edge path between preserving the exchange values of past capital investments in the built environment and destroying the value of these investments in order to open up fresh room for accumulation. Under capitalism there is, then, a perpetual struggle in which capital builds a physical landscape appropriate to its own condition at a particular moment in time, only to have to destroy it, usually in the course of a crisis, at a subsequent point in time. The temporal and geographical ebb and flow of investment in the built environment can be understood only in terms of such a process. The effects of the internal contradictions of capitalism, when projected into the specific context of fixed and immobile investment in the built environment, are thus writ large in the historical geography of the landscape that results.

V. CLASS STRUGGLE, ACCUMULATION, AND THE URBAN PROCESS UNDER CAPITALISM

What, then, of overt class struggle – the resistance that the working class collectively offers to the violence that the capitalist form of accumulation inevitably inflicts upon it? This resistance, once it becomes more than merely nominal, must surely affect the urban process under capitalism in definite ways. We must, therefore, seek to incorporate some understanding of it into any analysis of the urban process under capitalism. By switching our window on the world – from the contradictory laws of accumulation to the overt class struggle of the working class against the effects of those laws – we can see rather different aspects of the same process with greater clarity. In the space that follows I shall try to illustrate the complementarity of the two viewpoints.

In one sense, class struggle is very easy to write about because there is no theory of it, only concrete social practices in specific social settings. But this immediately places upon us the obligation to understand history if we are to understand how class struggle has entered into the urban process. Plainly I cannot write this history in a few pages, so I shall confine myself to a consideration of the contextual conditions of class struggle and the nature of the bourgeois responses. The latter are governed by the laws of accumulation because accumulation always remains the means whereby the capitalist class reproduces itself as well as its domination over labor.

The central point of tension between capital and labor lies in the workplace and is expressed in struggles over the work process and the wage rate. These struggles take place in a context. The nature of the demands, the capacity of workers to organize, and the resolution with which the struggles are waged depend a great deal upon the contextual conditions. The law (property rights, contract, combination and association, etc.), together with the power of the capitalist class to enforce its will through the use of state power, is obviously fundamental, as any casual reading of labor history will abundantly illustrate. What specifically interests me here, however, is the process of reproduction of labor power in relation to class struggle in the workplace.

Consider, first, the quantitative aspects of labor power in relation to the needs of capitalist accumulation. The greater the labor surplus and the more rapid its rate of expansion, the easier it is for capital to control the struggle in the workplace. The principle of the industrial reserve army under capitalism is one of Marx's most telling insights. Migrations of labor and capital as well as the various mobilization processes by means of which "unused" elements in the population are drawn into the workforce are manifestations of this basic need for a relative surplus population. But we also have to consider the costs of reproduction of labor power at a standard of living which reflects a whole host of cultural, historical, moral, and environmental considerations. A change in these costs or in the definition of the standard of living has obvious implications for real-wage demands and for the total wage bill of the capitalist class. The size of the internal market formed by the purchasing power of the working class is not irrelevant to accumulation either. Consequently, the consumption habits of the workers are of considerable direct and indirect interest to the capitalist class.

But we should also consider a whole host of qualitative aspects of labor power encompassing not only skills and training but attitudes of mind, levels of compliance, the pervasiveness of the work ethic and of "possessive individualism," and the variety of fragmentations within the labor force which derive from the division of labor and occupational roles, as well as from older fragmentations along racial, religious, and ethnic lines. The ability and urge of workers to organize along class lines depends upon the creation and maintenance of a sense of class consciousness and class solidarity in spite of these fragmentations. The struggle to overcome these fragmentations in the face of divide-and-conquer tactics often adopted by the capitalists is fundamental to understanding the dynamics of class struggle in the workplace.

This leads us to the notion of *displaced* class struggle, by which I mean class struggle that has its origin in the work process but that ramifies and reverberates throughout all aspects of the system of relations which capitalism establishes. We can trace these reverberations to every corner of the social totality and certainly see them at work in the flows of capital between the different circuits. For example, if productivity fails to rise in the workplace, then, perhaps judicious investment in human capital (education), in co-optation (homeownership for the working class), in integration (industrial democracy), in persuasion (ideological indoctrination), or in repression might yield better results in the long run. Consider, as an example, the struggles around public education. In *Hard Times,* Dickens constructs a brilliant satirical counterpoint between the factory system and the educational, philanthropic, and religious institutions designed to cultivate habits of mind amongst the working class conducive to the workings of the factory system, while elsewhere he has that archetypal bourgeois, Mr. Dombey, remark that public education is a most excellent thing provided it teaches the common people their proper place in the world. Public education as a right has long been a basic working-class demand. The bourgeoisie at some point grasped that public education could be mobilized against the interests of the working class. The struggle over social services in general is not merely over their provision but over the very nature of what is provided. A national health care system that defines ill health as inability to go to work (to produce surplus value) is very different indeed from one dedicated to the total mental and physical well-being of the individual in a given physical and social context.

The socialization and training of labor – the management of human capital – cannot be left to chance. Capital therefore reaches out to dominate the living process – the reproduction of labor power – and it does so because it must. The links and relations here are intricate and difficult to unravel. Next I consider various facets of activity within the dwelling place as examples of displaced class struggle.

Some Remarks on the Housing Question

The demand for adequate shelter is clearly high on the list of priorities from the standpoint of the working class. Capital is also interested in commodity production for the consumption fund, provided this presents sufficient opportunities for accumulation. The broad lines of class struggle around the housing question have had a major impact upon the urban process. We can trace some of the links back to the workplace directly. The agglomeration and concentration of production posed an immediate quantitative problem for housing workers in the right locations – a problem that the capitalist initially sought to resolve by the production of company housing but that thereafter was left to the market system. The cost of shelter is an important item in the cost of labor power. The more workers have the capacity to press home wage demands, the more capital becomes concerned about the cost of shelter. But housing is more than just shelter. To begin with, the whole structure of consumption in general relates to the form that housing provision takes. The dilemmas of potential overaccumulation which faced the United States in 1945 were in part resolved by the creation of a whole new life style through the rapid proliferation of the suburbanization process. Furthermore, the social unrest of the 1930s pushed the bourgeoisie to adopt a policy of individual homeownership for the more affluent workers as a means to ensure social stability. This solution had the added advantage of opening up the housing sector as a means for rapid accumulation through commodity production. So successful was this solution that the housing sector became a Keynesian "contra-cyclical" regulator of the accumulation process as a whole, at least until the *débâcle* of 1973. The lines of class struggle in France were markedly different (see Houdeville 1969). With a peasant sector to ensure social stability in the form of small-scale private property ownership, the housing problem was seen politically mainly in terms of costs. The rent control of the interwar years reduced housing costs but curtailed housing as a field for commodity production with all kinds of subsequent effects on the scarcity and quality of housing provision. Only after 1958 did the housing sector open up as a field for investment and accumulation – and this under government stimulus. Much of what has happened in the housing field and the shape of the "urban" that has resulted can be explained only in terms of these various forms of class struggle.

The "Moral Influence" of Suburbanization as an Antidote to Class Struggle

My second example is even more complex. Consider, in its broad outlines, the history of the bourgeois response to acute threats of civil strife, which are often associated with marked concentrations of the working class and the unemployed in space. The revolutions of 1848 across Europe, the Paris Commune of 1871, the urban violence that accompanied the great railroad strikes of 1877 in the United States and the Haymarket incident of 1886 in Chicago, clearly demonstrated the revolutionary dangers associated with the high concentration of the "dangerous classes" in certain areas. The bourgeois response was in part characterized by a policy of dispersal so that the poor and the working class could be subjected to what nineteenth-century urban reformers on both sides of the Atlantic called the "moral influence" of the suburbs. Cheap suburban land, housing, and transportation were all a part of this solution entailing, as a consequence, a certain form and volume of investment in the built environment on the part of the bourgeoisie. To the degree that this policy was necessary, it had an important impact upon the shape of both British and American cities. And what was the bourgeois response to the urban riots of the 1960s in the ghettos of the United States? Open up the suburbs, promote low-income and black homeownership, improve access via the transport system . . . the parallels are remarkable.

The Doctrine of "Community Improvement" and Its Contradictions

The alternative to dispersal lies in the application of doctrines of community improvement. As early as 1812, the Reverend Thomas Chalmers proposed to mobilize the "spirit of community" as an antidote to the class consciousness and its associated threat of revolutionary violence then engulfing the rapidly growing proletariat in British cities. In Chalmers's hands this mainly meant the use of the church and other community institutions as weapons of ideological control, searching to promote a doctrine of community harmony in the face of the realities of class struggle. But in the hands of the civic, urban, and "moral" reformers of the late nineteenth century (in countries as diverse as Britain, France, and the United States) it meant a real effort to improve the qualities of life for at least the respectable working class if not for the urban poor. In the hands of sociologists like Le Play and the founders of the Chicago School, the religious imperative was subtly buried within seemingly neutral principles of scientific enquiry that also suggested modes of social action to counter the threat of social unrest. From the urban reformers like Joseph Chamberlain of Birmingham and the "progressives" of the United States, to the architects of the "great society" programs in the 1960s in the United States, we find a continuous thread of bourgeois response to a structural problem that just will not disappear.

But the "principle of community" is not a bourgeois invention. It also has its authentic working-class counterpart as a defensive and even an offensive weapon in class struggle. The conditions of life in the community are of great import to the working class, and they can therefore become a focus of struggle which can assume a certain relative autonomy from that waged in the factory. The institutions of community can be captured and put to work for working-class ends. The church in the early years of the industrial revolution was on occasion mobilized at the local level in the interests of the working class much as it also became a focus for the black liberation movement in the United States in the 1960s and is a mobilization point for class struggle in the Basque country of Spain. The principle of community can then become a springboard for class action rather than an antidote to class struggle. Indeed, we can argue that the definition of community as well as the command of its institutions is one of the stakes in class struggle in capitalist society. This struggle can break open into innumerable dimensions of conflict, pitting one element within the bourgeoisie against another and various fragments of the working class against others as the principles of "turf" and "community autonomy" become an essential part of life in capitalist society. The bourgeoisie has frequently sought to divide and rule but just as frequently has found itself caught in the harvest of contradictions it has helped to sow. We find bourgeois suburbanites resisting the further accumulation of capital in the built environment, and individual communities in competition for development producing a grossly inefficient and irrational spatial order even from the standpoint of capital at the same time as they incur levels of indebtedness which threaten financial stability (the dramatic fiscal difficulties of New York City, 1973–75, is typical of the historical experience of the United States). We find also civil disorder within the urban process escalating out of control as ethnic, religious, and racial tensions take on their own dynamic in partial response to bourgeois promptings (the use of ethnic and racial differences by the bourgeoisie to split organization in the workplace has a long and ignoble history in the United States in particular).

Working-Class Resistance and the Circulation of Capital

The strategies of dispersal, community improvement, and community competition, arising as they do out of the bourgeois response to class antagonisms, are fundamental to understanding the material history of the urban process under capitalism. And they are not without their implications for the circulation of capital either. The direct victories and concessions won by the working class have their impacts. But at this point we come back to the principles of accumulation, because if the capitalist class is to reproduce itself and its domination over labor it must effectively render whatever concessions labor wins from it consistent with the rules governing the productivity of investments under capitalist accumulation. Investments may switch from one sphere to another in response to class struggle to the degree that the rules for the accumulation of capital are observed. Investment in working-class housing or in a national health service can thus be transformed into a vehicle for accumulation via commodity production for these sectors. Class struggle can, then, provoke switching crises, the outcome of which can change the structure of investment flows to the advantage of the working class. But those demands that lie within the economic possibilities of accumulation as a whole can in the end be conceded by the capitalist class without loss. Only when class struggle pushes the system beyond its own internal potentialities are the accumulation of capital and the reproduction of the capitalist class called into question. How the bourgeoisie responds to such a situation depends on the possibilities open to it. For example, if capital can switch geographically to pastures where the working class is more compliant, then it may seek to escape the consequences of heightened class struggle in this way. Otherwise it must invest in economic, political, and physical repression or simply fall before the working-class onslaught.

Class struggle thus plays its part in shaping the flows of capital between spheres and regions. The timing of investments in the built environment of Paris, for example, is characterized by deep troughs in the years of revolutionary violence – 1830, 1848, 1871 (see fig. 5). At first sight the rhythm appears to be dictated by purely political events, yet the typical fifteen- to twenty-five-year rhythm works just as well here as it does in other countries where political agitation was much less remarkable. The dynamics of class struggle are not immune to influences stemming from the rhythms of capitalist accumulation, of course, but it would be too simplistic to interpret the political events in Paris solely in these terms (see *The Urbanization of Consciousness*, chap. 3). What seems so extraordinary is that the overall rhythms of accumulation remain broadly intact in spite of the variations in the intensity of working-class struggle.

But if we think it through, this is not, after all, so extraordinary. We still live in a capitalist society. And if that society has survived, then it must have done so by imposing those laws of accumulation whereby it reproduces itself. To put it this way is not to diminish working-class resistance but to show that a struggle to abolish the wages system and the domination of capital over labor must necessarily look to the day when the capitalist laws of accumulation are themselves relegated to the history books. And until that day, the capitalist laws of accumulation, replete with all of their internal contradictions, must necessarily remain the guiding force in our history.

2

The Geography of Capitalist Accumulation: Toward a Reconstruction of the Marxian Theory

The need of a constantly expanding market for its products chases the bourgeoisie over the whole surface of the globe. It must nestle everywhere, settle everywhere, establish connexions everywhere. . . . All old established national industries have been destroyed or are daily being destroyed. They are dislodged by new industries, whose introduction becomes a life and death question for all civilized nations, by industries that no longer work up indigenous raw material but raw material drawn from the remotest zones; industries whose products are consumed, not only at home, but in every quarter of the globe. In place of the old wants, satisfied by the productions of the country, we find new wants, requiring for their satisfaction the products of distant lands and climes. In place of the old local and national seclusion and self-sufficiency, we have intercourse in every direction, universal inter-dependence of nations. (*The Communist Manifesto,* 46–47)

The geographical dimension of Marx's theory of the accumulation of capital has for too long been ignored. This is, in part, Marx's own fault, since, in spite of the dramatic depiction of the global conquests of the bourgeoisie in *The Communist Manifesto,* his writings on the topic are fragmentary, casually sketched, and unsystematic. His intention was, apparently, not to leave matters in such an unordered state. But projected books on the state, the world market, and crisis formation were never given even serious consideration. Careful scrutiny of his works reveals, however, a scaffold of thought on the subject that can bear the weight of substantive theorizing and historical interpretation. My purpose is to give more explicit shape and substance to that scaffold and so to lay the basis for more profound theorizing about the spatial dynamics of accumulation. This will, I hope, help us elucidate and interpret the actual historical geography of capitalism.

The importance of such a step scarcely needs stressing. Phenomena as diverse as urbanization, uneven geographical development, interregional interdependence and competititon, restructurings of the regional and international division of labor, the territoriality of community and state functions, imperialism, and the geopolitical struggles that flow therefrom all stand to be elucidated and incorporated into the grand corpus of theory that Marx bequeathed us. The trick is to unravel the relation between the temporal dynamics of accumulation of capital and the production of new spatial configurations of production, exchange, and consumption.

But the path to such an understanding is littered with all manner of obstacles. The theory that Marx did produce usually treats capitalism as a closed system. External space relations and internal spatial organization apparently play no role in shaping temporal dynamics. And most Marxists have followed Marx in this and so have produced an extraordinary bias within that tradition against any explicit theorization of space and space relations. How, then, can we rectify this omission and insert space and geography back into the argument? In what follows I shall insist that space and geography not be treated as afterthoughts, as mere appendages to already achieved theory. There is more to the problem than merely showing how capitalism shapes spatial organization, how it produces and continuously revolutionizes its geographical landscape of production, exchange, and consumption. I shall argue that space relations and geographical phenomena are fundamental material attributes that have to be present at the very beginning of the analysis and that the forms they assume are not neutral with respect to the possible paths of temporal development. They are to be construed, in short, as fundamental and "active moments" within the contradictory dynamics of capitalism. My grounds for this insistence are twofold.

First, I interpret Marx's method not as seeking firm and immutable conceptual building blocks from which to derive conclusions but as a process that moves dialectically and that, at each new phase, extends, revises, and expands our interpretations of the basic categories with which we commenced our investigations. The investigation of the dynamics of crisis formation, of the circulation of fixed capital, of the operations of the credit system, for example, all lead to significant reformulations of basic concepts like "use value" and "value" (Harvey 1982). The same proposition holds when we consider spatial dynamics and geographical phenomena. Withholding consideration of space and geography at the outset, as Marx tends to do, has no necessarily deleterious effects upon our final understandings – provided, therefore, that we take seriously the admittedly difficult task of reformulating our conceptual apparatus as we go.

Second, there is abundant evidence within Marx's own often fragmentary comments that this is exactly what he thought to do. To begin with, the first chapters of *Capital* incorporate several spatial concepts (community, place, world market, etc.), while spatial images abound (value is understood with the help of a geometric illustration and is then described as "crystals" – an image in which the transformation of spatial form and content are integral to each other – of some social substance). There is a sense, then, in which Marx does not exclude such issues even at the outset of his analysis. Furthermore, the phrases he uses throughout frequently indicate a close connection between spatial and geographical phenomena on the one hand and the basic conceptual apparatus on the other. In what follows we shall hit upon abundant examples of such a connection. But the point is important enough to warrant some preliminary demonstration.

Early in *Capital* (1:86–89), Marx notes that "money is a *crystal*" formed "of necessity" through "the historical progress and *extension*" of exchange which "develops the contrast, latent in commodities, between use value and value." It then follows that "the value of commodities more and more expands into an embodiment of human labour in the abstract" in "proportion as exchange *bursts its local bonds.*" This is a familiar theme in Marx. It says that the growth of trade across the world market is fundamental to the use value/value distinction as well as to the distinction between concrete and abstract labor. To the degree that the latter distinction is "the pivot upon which a clear comprehension of Political Economy turns," who can doubt that the study of the geographical integration of market exchange and the circulation of capital, of changing space relations, has much to say about the interpretation to be put on value itself? And this is no isolated instance either. We find Marx arguing, for example, that transportation over space is "productive of value," that the capacity to overcome spatial barriers belongs to "the productive forces," that the detail and social division of labor depends upon the agglomeration of laborers and the concentration of productive force in space, that differentials in labor productivity have a "basis" in natural differentiations, that the value of labor power varies according to geographical circumstances, and the like. Whenever spatial and geographical phenomena are introduced, the fundamental conceptual apparatus is usually not far behind. Such phenomena have therefore to be accorded a fundamental position in the overall theory.

Our task, then, is to bring spatial relations and geographical phenomena explicitly into the main corpus of Marxian theory and to trace the effects of such an insertion upon our interpretations of fundamental concepts. The first step is to search among the clues liberally sprinkled in Marx's own writings to get a sense of directions to take and paths to explore. The harder we push this kind of research, the closer we shall come to creating a theory with which to understand the central dynamics of capitalism's historical geography. And that, surely, is no mean agenda for research.

I. TRANSPORTATION RELATIONS, SPATIAL INTEGRATION, AND THE ANNIHILATION OF SPACE BY TIME

The *circulation of capital* in its standard form can be defined as a continuous process: money is used to buy commodities (labor power and means of production), which, when transformed through production, allows a fresh commodity to be thrown upon the market in exchange for the initial money outlay plus a profit. The *circulation of commodities,* however, refers simply to the patterns of market exchange of commodities. While there can be market exchange of commodities without the circulation of capital, the latter presupposes the former. For purposes of analysis, therefore, we can begin by isolating the exchange of commodities as a single transitional moment in the overall circulation of capital. By analyzing the conditions of the circulation of commodities, we can prepare the way for a more thorough understanding of the circulation of capital in space.

When mediated by money, the circulation of commodities "bursts through all restrictions as to time, place, and individuals" (*Capital* 1:113). Selling in one place and buying in another while holding money in between becomes a normal social act. When aggregated, the innumerable acts of buying and selling define the circulation processes of both money and commodities. These processes entail costs of two sorts (*Capital* 2: chap. 6). What Marx calls the "*faux frais*" of circulation are regarded as necessary but unproductive costs, necessary deductions from surplus value created in production. These include costs of circulation such as storage, bookkeeping, and the labor expended and profit extracted from retailing, wholesaling, banking, legal and financial services, and the like. These costs contrast with the expenditure of labor power to move commodities, money, and information from one place to another.

An analysis of the separation between buying and selling in space leads directly, therefore, to a consideration of the role of transport and communications in the circulation of commodities and money, and, hence, of capital. Marx has a fair amount to say on this topic. The industry that "sells change of location" as its product, he argues, is directly productive of value because "economically considered, the spatial condition, the bringing of the product to market, belongs to the production process itself. The product is really finished only when it is on the market" (*Grundrisse,* 533–34; *Capital* 2:150). This means that capital can be productively invested to enhance the circulation of commodities across space. However, the industry has its own peculiar laws of production and realization, because transportation is itself produced and consumed simultaneously at the moment of its use, while it also typically relies heavily upon fixed capital (roadbeds, terminals, rolling stock, and the like). Although there is potential here for direct surplus value production, there are good reasons for capitalists not to engage in its production except under certain favorable circumstances. The state is often, therefore, very active in this sphere of production (*Grundrisse,* 533–34).

Any reduction in the cost of transportation is important, Marx argues, because "the expansion of the market and the exchangeability of the product," as well as the prices of both raw materials and finished goods, are correspondingly affected (loc cit.). The ability to draw in raw materials over long distances and to dispatch products to distant markets is obviously affected by these costs. Such cost reductions depend upon the production of "improved, cheaper and more rapid transportation" (*Capital* 2:142). Viewed from the standpoint of production in general, therefore, "the reduction of the costs of real circulation [in space] belongs to the development of the forces of production by capital" (*Grundrisse,* 533–34).

Put in the context of Marx's general proposition of the impulsion, under capitalism, to perpetually revolutionize the productive forces, this implies an inevitable trend toward perpetual improvements in transportation and communications. Marx provides some hints as to how pressure is brought to bear to achieve such improvements. "The revolution in the modes of production of industry and agriculture made necessary a revolution . . . in the means of communication and transport," so that they "became gradually adapted to the modes of production of mechanical industry, by the creation of a system of river steamers, railways, ocean steamers and telegraphs" (*Capital* 1:384).

Elsewhere, he advances the following general proposition: "the more production comes to rest on exchange value, hence on exchange, the more important do the physical conditions of exchange – the means of communication and transport – become for the costs of circulation. Capital by its nature drives beyond every spatial barrier. Thus the creation of the physical conditions of exchange . . . becomes an extraordinary necessity for it" (*Grundrisse,* 524). The consequent reduction in transport costs opens up fresh pastures for the circulation of commodities and, hence, of capital. "The direct product can be realized in distant markets in mass quantities," and new "spheres of realization for labor driven by capital" can be opened up (loc cit.).

But the movement of commodities over greater distances, albeit at lower cost, tends to increase the time taken up during circulation. The effect is to increase the turnover time of capital – defined as the production time plus the circulation time (*Capital* 2:248), unless there are compensating improvements in the speed of circulation. Since the longer the turnover time of a given capital the smaller its annual yield of surplus value, the speed of commodity circulation is just as important to the circulation of capital as the cost. Marx takes up this idea explicitly. Speeding up "the velocity of circulation of capital" in the spheres of both production and exchange contributes to the accumulation of capital. From the standpoint of the circulation of commodities, this means that "even spatial distance reduces itself to time: the important thing is not the market's distance in space, but the speed . . . with which it can be reached" (*Grundrisse,* 538). There is every incentive, therefore, to reduce the circulation time of commodities to a minimum (*Capital* 2:249). A dual need, to reduce both the cost and the time of movement, therefore arises out of the imperatives of accumulation of capital: "While capital must on one side strive to tear down every spatial barrier to intercourse, i.e., to exchange, and conquer the whole earth for its market, it strives on the other side to annihilate this space with time. . . . The more developed the capital . . . the more does it strive simultaneously for an even greater extension of the market, and for greater annihilation of space by time" (*Grundrisse,* 539).

The phrase "annihilation of space by time" is of great significance within Marx's thinking. It suggests that the circulation of capital makes time the fundamental dimension of human affairs. Under capitalism, after all, it is *socially necessary labor time* that forms the substance of value, *surplus labor time* that lies at the origin of profit, and the ratio of surplus labor time to *socially necessary turnover time* that defines the rate of profit and, ultimately, the average rate of interest (Harvey 1982, chaps. 9 and 10). Under capitalism, therefore, the meaning of space and the impulse to create new spatial configurations of human affairs can be understood only in relation to such temporal requirements. The phrase "annihilation of space by time" does not mean that the spatial dimension becomes irrelevant. It poses, rather, the question of how and by what means space can be used, organized, created, and dominated to fit the rather strict temporal requirements of the circulation of capital.

Consideration of that question leads Marx down a number of interesting paths. He argues, for example, that the continuity of flow across space and the regularity of delivery play significant roles in relation to turnover time – the reduction of reserve stocks and of inventories of all types reduces the quantity of capital necessarily kept idle within the overall turnover process. It follows that there is a strong need to organize the transport and communications system to guarantee regularity of delivery as well as speed and low cost (*Capital* 2:249–50).

But the temporal requirements of the circulation of capital prompt further important adjustments within the organization of capitalism to deal with the spatial barriers it encounters. Long-distance trade, because it separates production and consumption by a relatively long time interval, poses serious problems for the continuity of capital flow. Herein, in Marx's opinion, lies "one of the material bases" of the credit system (*Capital* 2:251–52). In the

Grundrisse (535) Marx develops this argument at greater length in language that renders explicit the relations between time, space, and the credit system under capitalism:

> Circulation appears as an essential process of capital. The production process cannot be begun anew before the transformation of the commodity into money. The *constant continuity* of this process, the unobstructed and fluid transition of value from one form into the other, or from one phase of the process into the next, appears as a fundamental condition for production based on capital to a much greater degree than for all earlier forms of production. [But] while the necessity of this continuity is given, its phases are separate in time and space. . . . It thus appears as a matter of chance . . . whether or not its essential condition, the continuity of the different processes which constitute its process as a whole, is actually brought about. The suspension of this chance element by capital itself is *credit*.

The credit system in effect permits money to circulate in space independently of the commodities for which that money is an equivalent. The circulation of credit on the world market then becomes one of the chief mechanisms for the annihilation of space by time and dramatically enhances the capacity to circulate commodities (and hence capital) across space. In the process a certain power devolves upon the money capitalists vis-à-vis industrialists, while the contradictions inherent in the credit system also take on specific geographical expression (*Capital* 3:338–613; Harvey 1982, chaps. 9 and 10). This is a matter that I shall take up again shortly.

The efficiency with which commodities can be circulated over space depends also upon the activities of the merchant capitalists. Marx here contrasts the historical role of the merchant – buying cheap in order to sell dear; mediating between geographically dispersed producers at low levels of development; accumulating capital through profiteering, robbery, and violence; and forming the world market (*Capital* 3: chap. 20) – with the position of the merchant under a purely capitalist mode of production. In the latter instance, Marx argues, the merchant's role is to lower the cost and speed up the circulation of commodities (and hence of capital) by specializing in the marketing function (*Capital* 3: chaps. 16–19). Profits are to be had out of the efficient performance of such a role. But, like the money capitalists, the position of the merchants in the overall circulation process of capital gives them a certain power vis-à-vis the industrial capitalists and provides all too frequent opportunity for free expression of their penchant for speculation, profiteering, cheating, and excessive accumulation. Nevertheless, to the degree that the heart of the modern form of capitalism is shaped by "the immanent necessity of this mode of production to produce on an ever-enlarged scale" (*Capital* 3:333), so the formation of the world market can no longer be attributed to the activities of the merchant but must be traced back to its origins in capitalist production.

The direct relaxation of spatial constraints through revolutions that reduce the cost and time of movement and improve its continuity and efficiency can therefore be supplemented by the increasing efficiency of organization of the credit and marketing system. The latter helps to annihilate space with time and so increases the capacity for spatial integration between geographically dispersed producers. The industrial capitalists, however, can achieve much the same effect through their organization of production, their location decisions, and their technological choices. Let us see how Marx deals with such a possibility.

The capacity to procure surplus value is linked to the physical productivity of the labor employed. Capitalists can here exploit those variations that have their origin in nature (*Capital* 1:178, 511–15). Superior locations can similarly be exploited in trade. Under the coercive laws of competition, therefore, we might reasonably expect the location of production to be increasingly sensitive to natural variation and locational advantage. Marx rejects such an idea without, however, denying the basis of human activity in nature and location. He insists, first of all, that fertility, productivity, and location are *social* determinations, subject to direct modification through human action and equally subject to reevaluation through changing technologies of production: "Capital with its accompanying relations springs from an economic soil that is the product of a long process of development. The productiveness of labour that serves as [capital's] foundation and starting point is a gift, not of Nature, but of history embracing thousands of centuries" (*Capital* 1:512). Fertility can be built up in the soil, relative locations altered by transport improvements, and new productive forces embedded by human labor in the land itself (*Capital* 3:619, 651, 781). Furthermore, the advantage of access to, say, a waterfall as a power source can be eliminated overnight by the advent of the steam engine. Marx is primarily interested in the way in which transformations of this sort liberate capitalist production from natural constraints and produce a humanly created "second nature" as the arena for human action. And if circumstances arise (and Marx concedes this was frequently the case in the agriculture of his time) where natural fertility and location continue to give permanent advantages to privileged producers, then the benefit could be taxed away as ground rent (see below, Chap. 6).

The location of production cannot, therefore, be interpreted as a mere response to natural conditions but as the outcome of a social process in which modifications of nature, of locational advantage, and of the labor process are linked. The persistence of spatial and resource endowment constraints has then to be interpreted as an effect internal to the logic of capitalist development rather than as something that resides in external nature. And this brings us back to the idea that one of the principles internal to the logic of capitalist organization of production is the annihilation of space by time and the reduction of spatial barriers.

For example, when capitalists seeking relative surplus value strive to mobilize and appropriate labor's powers of cooperation, they do so by concentrating activity within a relatively smaller space (*Capital* 1:329). The reorganization of the detail division of labor for the same purpose demands that processes that were once successive in time "go on side by side in space" simultaneously (*Capital* 1:344). The application of machinery and the rise of the factory system consolidate this tendency toward the spatial concentration of labor and productive forces within a restricted space. This same principle carries over to questions of interindustry linkages within the social division of labor. The agglomeration of production within a few large urban centers, the workshops of capitalist production, is a tendency inherent in the capitalist mode of production (*Capital* 1:532; *Grundrisse,* 587; *The Communist Manifesto,* 47–48). In all of these instances we see that the rational organization of production in space is fundamental to the reduction of turnover time and costs within the circulation process of capital.

The tendency toward the agglomeration of population and productive forces in large urban centers is reinforced by a number of other processes of considerable significance. Technological innovations that liberate industry from close dependence upon a particular and localized raw material or energy source permit greater concentration of production in urban centers. This was precisely the importance of the steam engine, which "permitted production to be concentrated in towns" because it was "of universal application and, relatively speaking, little affected by its choice of residence by local circumstances" (*Capital* 1:378). Improvements in the means of transportation also tend "in the direction of the already existing market, that is to say, towards the great centers of production and population, towards ports of export, etc. . . . These particularly great traffic facilities and the resultant acceleration of capital turnover . . . give rise to quicker concentration of both the centers of production and the markets" (*Capital* 2:250).

The resultant "concentration of masses of men and capital thus accelerated at certain points" is even further emphasized because "all branches of production which by the nature of their product are dependent mainly upon local consumption, such as breweries, are . . . developed to the greatest extent in the principal centers of population" (loc. cit.). What Marx in effect depicts are powerful cumulative forces making for the production of urbanization under capitalism. And he helps us see these forces as part and parcel of the general processes seeking the elimination of spatial barriers and the annihilation of space by time.

But this process also requires the agglomeration of laborers, the concentration of population, within the restricted space of urban centers. "The more rapidly capital accumulates in an industrial or commercial town, the more rapidly flows the stream of exploitable human material" (*Capital* 1:661). This flow can arise out of "the constant absorption of primitive and physically uncorrupted elements from the country," which presupposes the existence there of "a constant latent surplus population" that can be dislodged by primitive accumulation – enclosures or other violent means of expropriation from the land (*Capital* 1:269, 642). The importation of Irish laborers into the industrial and commercial centers of England was of particular interest to Marx, for it not only provided a necessary flow of surplus laborers but did so in a way that divided the working-class movement (*Selected Correspondence*, 236–37). In the absence of such migrations, the expansion of the labor force depended upon the "rapid renewal" and "absolute increase" of a laboring population through fundamental and distinctively urban transformations in the social conditions of reproduction of labor power – earlier marriages, employment opportunities for children that encouraged laborers to "accumulate" children as their only source of wealth, and so on. And in the event of labor shortage, technological change tended to produce a "floating" industrial reserve army concentrated "in the centres of modern industry" (*Capital* 1:641). Even under conditions of technologically induced high unemployment, the capitalists could still leave the reproduction of labor power to "the labourer's instincts of self-preservation and of propagation" (*Capital* 1:572). The accumulation of capital in space goes hand in hand with "the accumulation of misery, agony of toil, slavery, ignorance, brutality, mental degradation," while children are raised in "conditions of infamy" (*Capital* 1:645).

Limits obviously exist to the progressive concentration of productive forces and laboring populations in a few large urban centers, even though such agglomeration may help reduce turnover times and circulation costs. Such concentrations of human misery form breeding grounds for class consciousness and organization, while overcrowding both in the factory and in the living space can become a specific focus of social protest (cf. *The Communist Manifesto*). But capital does not wait upon the emergence of such problems to set in motion its own quest for dispersal. The tendency to create the world market is, after all, "given in the concept of capital itself." The creation of surplus value at "one point requires the creation of surplus value at *another* point," which means "the production of a constantly widening sphere of circulation" through complementary tendencies to create new points of production and exchange. The exploration of "all nature in order to discover new, useful qualities of things," as well as to gain access to raw materials, entails "universal exchange of the products of all alien climates and lands"

(*Grundrisse,* 407–10). The tendency toward agglomeration is partially offset, therefore, by an increasingly specialized "territorial division of labour which confines special branches of production to special districts of a country" coupled with the rise of a "new and international division of labour" responsive to the needs of modern industry (*Capital* 1:353, 445–51). And all of this is facilitated by new transport and credit systems that facilitate long-distance movement, reduce spatial barriers, and annihilate space by time.

The conception toward which Marx appears to be moving is of a geographical landscape beset by a pervasive tension between forces making for agglomeration in place and forces making for dispersal over space in the struggle to reduce turnover time and so gain surplus value. If there is any general structure to it all – and Marx is far from explicit on the point – it is that of a progressive concentration of forces of production (including labor power) at particular places together with rapid geographical expansion of market opportunities. With the accumulation of capital, Marx comments, "flows in space" increase remarkably; while the "market expands spatially, the periphery in relation to the center is circumscribed by a constantly expanding radius" (*Theories of Surplus Value,* pt. 3:288). Some sort of center-periphery relation, perhaps an echo of that original antithesis between town and country which lies at the origin of the social division of labor (*Capital* 1:352; *The German Ideology,* 69), appears almost certain to arise.

But such a structure is perpetually being recast in the restless quest for accumulation. The creation of absolute surplus value rests upon "the production of a constantly widening sphere of circulation," while the production of relative surplus value entails "quantitative expansion of existing consumption: . . . creation of new needs by propagating existing ones in a wide circle" and "production of new needs and discovery and creation of new use values" through "the exploration of the earth in all directions." Marx then goes on to integrate the rise of science, the definition of new social wants and needs, and the transformation of world culture into his general picture of the global transformations necessarily wrought through an expansionary capitalism powered by the impulsion of accumulation for accumulation's sake:

> Capital drives beyond national barriers and prejudices as much as beyond nature worship, as well as all traditional, confined, complacent, encrusted satisfactions of present needs, and reproductions of old ways of life. It is destructive towards all of this and constantly revolutionizes it, tearing down all the barriers which hem in the development of the forces of production, the expansion of needs, the all-sided development of production, and the exploitation and exchange of natural and mental forces." (*Grundrisse,* 407–10).

But to the degree that capitalist production "moves in contradictions which are constantly overcome but just as constantly posited," so we find contradictions internal to this overall expansionary dynamic. In particular, the search for "rational" geographical configurations of production and consumption runs up against the impulsion to revolutionize the productive forces in transport and communications. Expansion occurs in a context where transformations in the cost, speed, continuity, and efficiency of movement over space alter "the relative distances of places of production from the larger markets." This entails "the deterioration of old and the rise of new centres of production" and a perpetual "shifting and relocation of places of production and of markets as a result of the changes in their relative positions" (*Capital* 2:249–50). This instability is exacerbated by processes of technological and organizational change that either liberate production from specific locational requirements (access to particular raw material or energy supplies, dependence upon particular labor skills) or confirm the trend toward increasing specialization within a territorial division of labor. And the shifting physical and social capacity of the laborers to migrate (on a temporary or permanent basis) also enters into the picture (Results of the Immediate Process of Production, 1013–14).

The shifts in spatial configurations produced by such processes become problematic to the degree that capitalism requires fixed and immobile infrastructures, tied down as specific use values in particular places, to facilitate production, exchange, transportation, and consumption. Capitalism, after all, "establishes its residence on the land itself and the seemingly solid presuppositions given by nature [appear] in landed property as merely posited by industry" (*Grundrisse,* 740). The value embodied in such use values cannot be moved without being destroyed. Capital thus must represent itself in the form of a physical landscape created in its own image, as use values created through human labor and embedded in the land to facilitate the further accumulation of capital. The produced geographical landscape constituted by fixed and immobile capital is both the crowning glory of past capitalist development and a prison that inhibits the further progress of accumulation precisely because it creates spatial barriers where there were none before. The very production of this landscape, so vital to accumulation, is in the end antithetical to the tearing down of spatial barriers and the annihilation of space by time.

This contradiction grows with increasing dependence upon fixed capital (machinery, plant, physical infrastructures of all kinds). The problem arises because with "fixed capital the value is imprisoned within a specific use value" (*Grundrisse,* 728), while the degree of fixity increases with durability, other things being equal (*Capital* 2:160). Marx describes the conditions governing the circulation of fixed capital in the following terms: "The value of fixed capital is reproduced only insofar as it is used up in the production process. Through disuse it loses its value without its value passing on to the product. Hence the greater the scale on which fixed capital develops . . . the more does the continuity of the production process or the constant flow of reproduction become an externally compelling condition for the mode of production founded on capital" (*Grundrisse,* 703). The employment of fixed and immobile capital, in short, exerts a strong claim upon the future circulation of capital and the future deployment of labor power. Until the capital invested in such assets is amortized through use, capital and labor power are constrained geographically to patterns of circulation that help realize the value embodied in all improvements "sunk in the soil . . . every form in which the product of industry is welded fast to the surface of the earth" (*Grundrisse,* 739–40; Harvey 1982, chap. 8).

Capitalist development has to negotiate a knife-edge path between preserving the values of past capital investments embodied in the land and destroying them in order to open up fresh geographical space for accumulation. A perpetual struggle ensues in which physical landscapes appropriate to capitalism's requirements are produced at a particular moment in time only to be disrupted and destroyed, usually in the course of a crisis, at a subsequent point in time.

This contradiction hides an irony that is nowhere more apparent than in the transport industry itself. The elimination of spatial barriers and the annihilation of space by time requires "a growth of that portion of social wealth which, instead of serving as direct means of production, is invested in means of transportation and communication and in the fixed and circulating capital required for their operation" (*Capital* 2:351). In other words, the production of a fixed configuration of space (e.g., the rail, road, and port systems) is the only means open to capital to overcome space. At some point the impulsion to overcome space must render the initial investments obsolete and redundant, perhaps well before the value embodied in them has been realized through use.

The location theory in Marx (if we may call it that) is not much more specific than this (although there is much of peripheral interest in his analyses of rent and fixed capital formation – see Harvey 1982, chaps. 8 and 11). The virtue of his fragmentary remarks lies not in their sophistication but in the vision they project of the role of the production and restless restructuring of geographical landscapes and space relations as active moments within the dynamics of capital accumulation. Revolutions in the productive forces embedded in the land, in the capacity to overcome space and annihilate space with time, are not afterthoughts to be added on in the final chapter of some analysis. They are fundamental because it is only through them that we can give flesh and meaning to those most pivotal of all Marxian categories, concrete and abstract labor.

This last point is of sufficient importance to warrant reflection. The expenditure of human labor power "in a special form and with a definite aim" to produce use values in a given place and time is, as Marx puts it, "an eternal nature-imposed necessity, without which there can be no material exchanges between man and Nature, and therefore no life." These different qualities of concrete labor are brought into relation with each other through exchange and, ultimately, through the circulation of capital. And that process of bringing different concrete labor activities into a general social relation gives that same labor process abstract qualities tied to value as socially necessary labor time, the labor time "required to produce an article under the normal conditions of production, and with the average degree of skill and intensity prevalent at the time" (*Capital* 1:39–46). The "normal conditions" and "average skill and intensity" cannot be specified, however, except with reference to a given space of exchange and capital circulation. The processes of formation of the world market, of spatial integration, of the international and territorial division of labor, of the geographical concentration of production (labor power and productive force), are therefore fundamental to understanding how a concrete labor process acquires abstract, universal qualities. For the geographer this must surely be one of Marx's most profound insights. For it not only puts the study of space relations and geographical differentiation into the heart of Marx's theorizing but it also points the way to a solution of the problem that has for so long bedeviled the geographical imagination: how to make universal generalizations about the evident unique particularities of space. The answer lies, of course, not in philosophical speculation but in a study of exactly how the processes of capital circulation bring the unique qualities of human action in given places and times into a framework of universal generality. And that, presumably, was exactly what Marx meant by that stunning conception that bears repeating: "Abstract wealth, value, money, hence abstract labour, develop in the measure that concrete labour becomes a totality of different modes of labour embracing the world market."

II. FOREIGN TRADE

While some of Marx's scattered comments on foreign trade (he never completed a projected work on the world market) can be interpreted as logical extensions of his views on location and space relations, the focus is more on how the history and dynamics of capital accumulation were and are expressed through preexisting geographical structures – the nation state in particular – rather than on the processes that give rise to spatial configurations in the first place. By accepting the fiction of capital accumulation as primarily a national affair, Marx was, of course, making concessions to a long line of thought that passed from the mercantilists via the physiocrats to Adam Smith and the

Ricardian doctrine of comparative advantage. The strength of that tradition in political economy inexorably drew Marx to a critique of some of its fundamental propositions and a partial acceptance of others. And if the picture he portrays appears somewhat at odds with that which we have previously outlined, it is no less legitimate for all that. It is, as it were, the world of geographical interactions seen from a rather different window. A full understanding of Marx's views must rest on a synthesis of two rather disparate but both equally legitimate perspectives on the geography of capitalist accumulation.

Marx sees the development of foreign trade, the formation of the world market, and the rise of capitalism as integrally related within a process in which consequences at one stage become preconditions for the next. The drive to overcome spatial barriers, for example, presages the absorption, dissolution, or transformation of all noncapitalist modes of production ultimately under the homogenizing force of the circulation of capital. Monetization, commodity exchange, and, finally, the imposition of capitalist relations of production represent various steps in such a process.

The mere penetration of the money form, he declares, has a "dissolving" influence on the isolated community and draws "new continents into the metabolism of circulation" (*Grundrisse,* 224–25). Capital can then be accumulated directly from that "metabolism of circulation" once it is established. The towns accumulate use values and hence values from the countryside, while merchants' capital, as a historically prior form of organization to producers' capital,

> appropriates an overwhelming portion of the surplus product partly as a mediator between communities which still substantially produce for use values . . . and partly because under these earlier modes of production the principal owners of the surplus product with whom the merchant dealt, namely, the slave owner, the feudal lord, and the state (for instance, the oriental despot), represent the consuming wealth and luxury which the merchant seeks to trap. . . . Merchant's capital, when it holds a position of dominance, stands everywhere for a system of robbery so that its development among the trading nations of old and modern times is always directly connected with plundering, piracy, kidnapping, slavery and colonial conquest. . . . The development of merchant's capital gives rise everywhere to the tendency towards production of exchange values. Commerce, therefore, has a more or less dissolving influence everywhere on the producing organization which it finds at hand and whose different forms are mainly carried on with a view to use value. (*Capital* 3:331–32)

Merchants' capital played a crucial role in the redistribution of wealth and power from the countryside to the city or from the whole world to a few dominant capitalist nations. But all this changed when merchants' capital became subservient to industrial capital and so was forced to respect the rules of fair exchange (*Capital* 1:165). The merchant then became the mere agent who imposed more basic forms of capitalist domination. For example:

The cheapness of the articles produced by machinery, and the improved means of transport and communication furnish weapons for conquering foreign markets. By ruining handicraft production in other countries, machinery forcibly converts them into fields for the supply of its raw material. In this way, East India was compelled to produce cotton, wool, hemp, jute and indigo for Great Britain. By constantly making a part of the hands "supernumerary," modern industry, in all countries where it has taken root, gives spur to emigration and to the colonization of foreign lands, which are thereby converted into settlements for growing the raw material of the mother country; just as Australia, for example, was converted into a colony for growing wool. A new and international division of labor, a division suited to the requirements of the chief centers of modern industry, springs up. (*Capital* 1:451)

The exact geography of this new international division of labor depends, however, upon a whole host of "special factors" and contradictory effects which make capitalist production's path to global domination peculiarly tortuous.

In the case of colonies, for example, Marx insists on what he sees as a key distinction:

There are colonies proper, such as the United States, Australia, etc. Here the mass of the farming colonists, although they bring with them a larger or smaller amount of capital from the motherland, are not *capitalists,* nor do they carry on *capitalist* production. They are more or less peasants who work themselves and whose main object, in the first place, is to produce their own livelihood. In the second type of colonies – plantations – where commercial speculations figure from the start and production is intended for the world market, the capitalist mode of production exists, although only in a formal sense, since the slavery of Negroes precludes free wage labor, which is the basis of capitalist production. But the business in which slaves are used is conducted by capitalists. (*Theories of Surplus Value,* pt. 2:302–3)

The two kinds of colony evolve very differently in relation to global processes of accumulation. Colonies of the second sort, no matter whether they are actively founded or fashioned out of a transformation of some precapitalist society (as in Eastern Europe), stand to be highly profitable, at least initially, because of the high rates of exploitation achievable through the reduction of necessities to a bare minimum. This tendency to transform necessities into luxuries

determines the whole social pattern of backward nations . . . which are associated with a world market based on capitalist production. No matter how large the surplus product they [the noncapitalist producers] extract from the surplus labor of their slaves in the simple form of cotton or corn, they can adhere to this simple undifferentiated labor because foreign trade enables them to convert these simple products into any kind of use value. (*Theories of Surplus Value,* pt. 3:243)

The inability to revolutionize the productive forces under such conditions of created underdevelopment is what in the long run renders such colonies vulnerable.

In contrast, colonies made up of small independent producers, trading some surplus into the market, are typically characterized by labor shortages and high wages (particularly where abundant, cheap land is available). Colonies of this sort are not so amenable to capitalist forms of exploitation and may even actively resist the penetration of the capitalist mode of production:

There the capitalist regime everywhere comes into collision with the resistance of the producer, who, as owner of his own conditions of labor, employs that labor to enrich himself instead of the capitalist. The contradiction of these two diametrically opposed economic systems manifests itself here practically in a struggle between them. Where the capitalist has at his back the power of the mother-country, he tries to clear out of his way by force the modes of production and appropriation based on the independent labor of the producer. (*Capital* 1:765)

Innumerable populist and radical movements spawned amongst settlers in frontier regions in the United States, Canada, Australia, and other countries testify to the importance of such a conflict. But since such colonies are shaped by the spin-off of surplus population backed by small quantities of capital from the main centers of accumulation, and since they also often form expanding markets for capitalist production, they, too, surely become integrated into a hegemonic capitalist mode of production. Thus the United States was being transformed in Marx's own time from an independent, largely noncapitalistic production system into a new center for capital accumulation. "Capitalistic production advances there with giant strides," Marx noted, "even though the lowering of wages and the dependence of the wage worker are yet far from being down to the European level" (*Capital* 1:773).

But there are other equally important "special factors" to be taken into account. Marx recognizes, for example, that "the productiveness of labor is fettered by physical conditions" and that differences in nature therefore form "a physical basis for the social division of labor." But he is equally emphatic that such differences represent possibilities only (and not unmodifiable by human action at that) because in the final analysis the productiveness of labor "is a gift, not of Nature, but of history embracing thousands of centuries" (*Capital* 1:512–14). Furthermore, to the degree that productivity is defined under capitalism to mean the capacity of the laborer to produce surplus value for the capitalist (*Capital* 1:509), so national and regional differences in the value of labor power become crucial:

> In the comparison of the wages in different nations, we must therefore take into account all the factors that determine changes in the amount of the value of labor-power; the price and extent of the prime necessaries of life as naturally and historically developed, the cost of training the laborers, the part played by the labor of women and children, the productiveness of labor, its extensive and intensive magnitude. (*Capital* 1:559)

The productivity of labor and the value of labor power, he concedes, can vary quite substantially, even within a country (*Wages, Price, and Profit,* 72–73). And capitalist production, far from eradicating such differences, can all too easily emphasize or even create them. "In proportion as capitalist production is developed in a country, in the same proportion do the national intensity and productivity of labor there rise above the international level" (*Capital* 1:560). The penetration of money relations and simple commodity exchange appears powerless to modify such a condition of uneven geographical development. And this has important implications:

> Capitals invested in foreign trade can yield a higher rate of profit, because, in the first place, there is competition with commodities produced in other countries with inferior production facilities, so that the more advanced country sells its goods above their value even though cheaper than the competing countries. . . . The favored country recovers more labor in exchange for less labor, though this difference, this excess is pocketed, as in any exchange between capital and labor, by a certain class. (*Capital* 3:238)

On this basis certain peculiarities can arise in the terms of trade between developed and underdeveloped societies, between centers and peripheries (*Theories of Surplus Value,* pt. 2:474–75). Furthermore, such differentials can persist. Countries may establish a monopoly over the production of particular commodities (*Capital*, 3:119), while other factors may also prevent any direct "levelling out of values by labor time and even the levelling out of cost prices by a general rate of profit" (*Theories of Surplus Value,* pt. 2:201). Even more startling is Marx's admission that

> here the law of value undergoes essential modification. The relationship between labor days of different countries may be similar to that existing between skilled, complex labor and unskilled, simple labor within a country. In this case the richer country exploits the poorer one, even when the latter gains by the exchange. (*Theories of Surplus Value,* pt. 3:105–6)

This assertion appears totally at odds with the main thrust of Marx's argument, that of inevitable global integration of capitalist production and exchange under a single law of value represented by universal money. It is, after all, "only in the markets of the world that money acquires to the full extent the character of the commodity whose bodily form is also the immediate social incarnation of human labor in the abstract" (*Capital* 1:142). Further,

> it is only foreign trade, the development of the market to a world market, which causes money to develop into world money and abstract labor into social labor. Abstract wealth, value, money, hence abstract labor, develop in the measure that concrete labor becomes a totality of different modes of labor embracing the world market. Capitalist production rests on the value or the transformation of the labor embodied in the product into social labor. But this is only possible on the basis of foreign trade and the world market. This is at once the pre-condition and the result of capitalist production. (*Theories of Surplus Value,* pt. 3:253)

Can we discern here a faint echo of that same contradiction that Marx makes more explicit in his consideration of location: that the elimination of geographical differentiations can be sought only through the active construction of fresh differentiations? It certainly seems as if specific material geographical structures mediate between the abstract aspects of labor (a social determination achieved through exchange on the world market) and labor's concrete qualities (the particularities of the labor process as undertaken by particular people in a particular place and time).

There are the kinds of "special factors" that make of foreign trade a very complex issue. These complexities do not derive from the failure of capitalist development to overcome the social and cultural barriers to its global hegemony (although these barriers can be exceedingly resistant and in some cases determinant). They stem, rather, from the inherent contradictions within the capitalist mode of production itself. Much of the complexity we encounter in the case of foreign trade must be interpreted, therefore, as global manifestations of the internal contradictions of capitalism. And underlying all such manifestations is the very real possibility that in the end capitalism creates the greatest barriers (geographical as well as social) to its own development. Deeper consideration of foreign trade in relation to the dynamics of capital accumulation can help establish exactly how this can be so.

III. THE GEOGRAPHY OF CRISIS FORMATION AND RESOLUTION: THE SEARCH FOR A "SPATIAL FIX" FOR CAPITALISM'S INTERNAL CONTRADICTIONS

Marx often specifically excludes foreign trade and geographical dimensions from his analysis of the dynamics of capitalist accumulation and crisis formation. Yet there are innumerable hints that he did not feel comfortable so doing. While he admits that "capitalist production does not exist at all without foreign commerce," he also argues that consideration of foreign trade merely confuses "without contributing any new elements of the problem [of accumulation], or of its solution" (*Capital* 2:470). While he accepts that foreign trade can counteract the tendency toward a falling rate of profit (and hence stave off crises), he hurriedly counters that this merely raises the rate of accumulation and so hastens the fall in the profit rate in the long run (*Capital* 3:237). While foreign trade evidently "transfers the contradictions [of capitalism] to a wider sphere and gives them greater latitude" (*Capital* 3:408), and while he also at one point anticipated a separate work on the world market and crises, he does not pay much attention in practice to such processes. For purposes of analysis, he argues, it is sufficient to proceed as if capitalism were a closed system: "in order to examine the object of our investigation in its integrity, free from all disturbing subsidiary circumstances, we must treat the world as one nation, and assume that capitalist production is everywhere established and has possessed itself of every branch of industry" (*Capital* 1:581).

Yet, in spite of these avowals, the very last chapter of the first volume of *Capital* takes up "the modern theory of colonization" and so opens up the whole question of foreign and colonial settlement and trade in a work that treats capitalism as a closed system. The position of the chapter is doubly odd in that it obscures what many would regard as a more natural culmination to Marx's argument in the penultimate chapter. There Marx announces, with a grand rhetorical flourish reminiscent of *The Communist Manifesto*, the death knell of capitalist private property and the inevitable expropriation of the expropriators (*Capital* 1:762–63).

The positioning and content of the final chapter reflects, I believe (cf. Harvey 1981), a desire on Marx's part to deal with a problem that Hegel had raised and left open in *The Philosophy of Right* (Hegel 1952, 149–52). Hegel there attributes the social unrest and evident instability of bourgeois civil society to the increasing polarization of the social classes – concentration of wealth in a few hands and the formation of "a rabble of paupers." This condition has its origin in the loss of property on the part of laborers and their consequent condemnation to an alienated existence, even though the laborers are the original source of all wealth. It also leads, for reasons that Hegel leaves obscure, to general crises of overproduction and underconsumption that civil society appears powerless to remedy. This "inner dialectic" forces civil society "to push beyond its own limits and seek markets, and so its necessary means of subsistence, in other lands that are either deficient in the goods it has overproduced, or else generally backward in industry." It must also found colonies and thereby permit a part of its unemployed population access to property and a return to an unalienated condition of existence while simultaneously supplying itself "with a new demand and field for its industry." Colonialism and imperialism are posited as necessary external resolutions to the internal contradictions of capitalism.

Much of Marx's *Capital* can be read as a transformation and materialist rendition of Hegel's idealist argument. Curiously, Marx avoided consideration of these questions in his *Critique of Hegel's Philosophy of Right*, but thirty years later, in an Afterword to *Capital* (1:19) admitted that his critique of Hegel had had a profound effect upon his mode of argumentation. The theme of increasing polarization of the social classes is writ large in Marx's "general law of capitalist accumulation." There Marx shows how and why capitalism reproduces "the capital relation on a progressive scale, more capitalists at this pole, more wage workers at that." Furthermore, the processes at work also produce a "relative surplus population," a "reserve army" of unemployed "set free" primarily through technological and organizational changes imposed by the capitalists. This reserve army helps drive wages down and control workers' movements. It is therefore a prime lever of accumulation. The effect, Marx observes in language deeply reminiscent of Hegel, is "the accumulation of wealth at one pole" and the "accumulation of misery, agony of toil, slavery, ignorance, brutality, mental degradation, at the opposite pole (*Capital* 1:630–45). Accumulation requires, then, that capitalists control both the demand for and supply of labor power. They must be able to create and maintain labor surpluses, either through the mobilization of "latent labor reserves" (women and children, peasants thrown off the land, etc.) or through the creation of technologically induced unemployment. Any threat to that control, Marx notes, is countered "by forcible means and State interference." In particular, capitalists must strive to check colonization processes that give laborers access to free land at some frontier (*Capital* 1:640). The colonial policies studied in the final chapter of *Capital* drive home Marx's argument with dramatic force – the powers of private property and the state were to be used to exclude laborers from easy access to free land in order to preserve a pool of surplus laborers for exploitation by capitalists.

There is, in this, a tacit answer to the question Hegel posed. If laborers can return to a genuinely unalienated life through migration to some frontier, then capitalist control over labor supply is undermined. Marx accepts the idea that higher wages and better conditions of life could exist in the United States precisely because of the laborer's access to a relatively open frontier. The survival of capitalism depends therefore upon the recreation of capitalist relations of private property and the associated power to appropriate the labor of others, even and especially at the open frontier. What Marx illustrates in his final chapter is the explicit bourgeois recognition of such a principle. He thereby signals a rejection of Hegel's proposition that there could be some "spatial fix" for capitalism's internal contradictions.

The same issue crops up again in Marx's consideration of crises. Polarization then takes the form of "unemployed capital at one pole, and unemployed worker population at the other." Can the formation of such crises be prevented or actual crises resolved through geographical expansion? Marx does not rule out the possibility that foreign trade can counteract the tendency toward a falling rate of profit in the short run (*Capital* 3:237–59). But how long is the short run? And if it extends over many generations, then what does this do to Marx's theory and its associated political practice of seeking revolutionary transformations in the heart of civil society in the here and now?

To answer this question we need a firmer interpretation of Marx's view of the "inner dialectics" of capitalism in crisis. This is not an uncontroversial matter, since rival interpretations of Marx's arguments abound. I shall work with a highly simplified version in which individual capitalists, locked in class struggle and coerced by competition, are forced into technological adjustments that destroy the prospects for balanced accumulation and so undermine the conditions of reproduction of both the capitalist and working classes. The end product of such a process is a condition of *overaccumulation of capital*, defined as an excess of capital in relation to opportunities to employ that capital profitably, and excess of labor power – widespread unemployment or underemployment (Harvey 1982, chap. 7). That capitalism should periodically produce such unemployable surpluses in the midst of such enormous human need was, for Marx, a tragic demonstration of the underlying irrationality of supposedly rational competitive market allocations. The "hidden hand" was by no means as benevolent as Adam Smith thought. That the only exit from such a state was the devaluation of both capital and labor power illustrates the destructive proclivities of bourgeois domination.

Is there, then, some way in which the surpluses of both capital and labor power generated in the course of crises can be successfully disposed of through geographical expansion? To answer that question we must first recognize that surpluses of capital can exist as money (a highly mobile form of capital), as commodities (variably mobile), or as idle productive capacity (hardly mobile at all). Marx's comments on the prospects for a spatial fix for capitalism's internal contradictions can then be assembled under three main headings.

External Markets and Underconsumption

In the course of a crisis,

> the English, for example, are forced to lend their capital to other countries in order to create a market for their commodities. Overproduction, the credit system, etc., are means by which capitalist production seeks to break through its own barriers and to produce over and above its own *limits*. . . . Hence crises arise, which simultaneously drive it onward and beyond [its own limits] and force it to put on seven-league boots, in order to reach a development of the productive forces which could be achieved only very slowly within its own limits. (*Theories of Surplus Value,* pt. 3:122; cf. *Grundrisse,* 416)

Such lending of excess money capital to a foreign country in order to allow the latter to buy up the excess commodities produced at home (so permitting the full employment of productive capacity and labor power at home) appears a neat way to deal with the overaccumulation problem. The difficulty is that it accelerates the development of the productive forces at home and so exacerbates the tendency toward overaccumulation. At the same time, it puts an undue debt burden on the foreign country which must at some point be paid. In the long run, therefore, the overaccumulation problem is merely intensified and spread out over a wider area. The collapse, when it comes, triggers an intricate sequence of events because of the gaps that exist between the imbalance of trade and the balance of payments. Marx describes a typical sequence this way:

> the crisis may first break out in England, the country which advances most of the credit and takes the least, because the balance of payments . . . which must be settled immediately, is *unfavorable,* even though the general balance of trade is *favorable*. . . . The crash in England initiated and accompanied by a gold drain, settles England's balance of payments. . . . Now comes the turn of some other country. . . .
>
> The balance of payments is in times of crisis unfavorable to every nation . . . but always to each country in succession, as in volley-firing. . . . It then becomes evident that all these nations have simultaneously over-exported (thus over-produced) and over-imported (thus over-traded), that prices were inflated in all of them, and credit stretched too far. And the same break-down takes place in all of them.

The costs of devaluation are then forced back onto the initiating region by,

> first, shipping away precious metals; then selling consigned commodities at low prices; exporting commodities to dispose of them or obtain money advances on them at home; increasing the rate of interest, recalling credit, depreciating securities, disposing of foreign securities, attracting foreign capital for investment in these

depreciated securities, and finally bankruptcy, which settles a mass of claims. (*Capital* 3:491–92, 517)

The sequence sounds dismally familiar. There is no prospect here, evidently, for a long-term spatial fix for capitalism's internal contradictions.

A more intriguing possibility arises with respect to trade with noncapitalist social formations. After all,

> when an industrial people, producing on the foundation of capital, such as the English, e.g. exchange with the Chinese, and absorb value . . . by drawing the latter within the sphere of circulation of capital, then one sees right away that the Chinese do not therefore need to produce as capitalists. (*Grundrisse,* 729)

Circumstances can then arise that make the development of capitalism "conditional on modes of production lying outside of its own stage of development." The degree of relief afforded thereby depends on the nature of the noncapitalist society and its capacity or willingness to integrate into the capitalist system in ways that absorb the excess capital. Crises, it transpires, can be checked only if the noncapitalist countries "consume and produce at a rate which suits the countries with capitalist production" (*Capital* 2:110; 3:257). And how can that be assured short of some form of political and economic domination? The current difficulties of East-West trade and the Polish debt form an excellent case in point. And even if such domination can be assured, the resolution is bound to be temporary. "You cannot continue to inundate a country with your manufactures," says Marx, "unless you enable it to give you some produce in return." Hence, "the more the [British] industrial interest became dependent on the Indian market, the more it felt the necessity of creating fresh productive powers in India" (*On Colonialism,* 52). And this broaches a whole new set of possible resolutions of the problem.

The Export of Capital for Production

Surplus capital can be lent abroad to create fresh productive powers in new regions. The higher rates of profit promised provide a "natural" incentive for such a flow and, if achieved, stand to increase the average rate of profit in the system as a whole (*Capital* 3:237, 256; *Theories of Surplus Value,* pt. 2:436–37). Crises are temporarily resolved. "Temporarily" because higher profits mean an increase in the mass of capital looking for profitable employment. The tendency toward overaccumulation is thereby exacerbated but now on an expanding geographical scale. The only escape lies in a continuous acceleration in the creation of fresh productive powers in new regions.

But when a particular civil society seeks to relieve its overaccumulation problem through the creation of fresh productive powers elsewhere, it thereby establishes a rival center of accumulation which, at some future point, will also run into problems of internal overaccumulation and seek refuge in some spatial fix. Marx thought he saw the first step down such a path as the British exported capital to India:

> When you have once introduced machinery into the locomotion of a country, which possesses iron and coals, you are unable to withhold it from its fabrication. You cannot maintain a net of railways over an immense country without introducing all those industrial processes necessary to meet the immediate and current wants of railway locomotion, and out of which there must grow the application of machinery to those branches of industry not immediately connected with the railways. The railway system will therefore become, in India, truly the forerunner of modern industry . . . [which] will dissolve the hereditary divisions of labor, upon which rest the Indian castes, those decisive impediments to Indian progress and Indian power. . . . The bourgeois period of history has to create the material basis of the new world. . . . Bourgeois industry and commerce create these material conditions of a new world in the same way that geological revolutions have created the surface of the earth. (*On Colonialism,* 85–87)

We here see that the capitalist imperative to overcome spatial barriers and annihilate space with time also necessitates revolutions in the productive forces in the spaces brought within the metabolism of global exchange. But the transition that Marx anticipated in India was blocked by a mixture of internal resistance to capitalist penetration and imperialist policies imposed by the British. The latter were, by and large, specifically geared to preventing the rise of India as a competitor, even to the point of destroying that country's industrial base in order to open up a market for British products. The transition was not blocked, however, in the United States, which was already becoming, as Marx noted in 1867, a new and independent center for capital accumulation.

These two contrasting examples hint at a more than trivial dilemma in the search for a spatial fix for overaccumulation through the export of capital. The dependent and constrained form of capitalist development in India avoided the problem of competition but quickly degenerated into a mere trading relation that, as we have seen, could do nothing for the long-run absorption of surplus capital and surplus labor power. The pace of growth and accumulation in India was not, in short, fast enough to absorb the ever-increasing surpluses emanating from Britain. The unconstrained development of capitalism in the United States, in contrast, absorbed far more excess British capital (and labor power) than India ever did, but the new productive forces created there posed a competitive threat to the initiating country.

While Marx does not make the point directly, we can quickly infer here the existence of a "Catch 22" of the following sort: if the new region is to absorb the surpluses from the home country effectively, then it must be allowed to develop freely into a full-fledged capitalist economy that is in turn bound to produce its own surpluses and so enter into competition with the home base. If, however, the new region develops in a constrained and dependent way, then competition with the home base is held in check but the rate of expansion is insufficient to absorb the burgeoning surpluses at home. Devaluation occurs no matter what. Unless, of course, still newer growth regions can be opened up. The effect, however, as Marx observes, is simply to spread the contradictions over ever wider geographical spheres and give them ever greater latitude of operation.

But, *nota bene,* capitalism can open up considerable breathing space for its own survival through this particular form of the spatial fix. It is rather as if, having sought to annihilate space with time, capitalism can then buy back time for itself out of the spaces it conquers. The geographical spread and intensification of capitalism is a long-drawn-out revolution accomplished over many years. While local, regional, and "switching" crises are normal grist for the working out of this process, the building of a truly global crisis of capitalism depends upon the exhaustion of possibilities for further revolutionary transformations along capitalist lines. And that depends upon the capacity to propagate new productive forces across the face of the earth, combined with the supply of fresh labor power open to exploitation as wage labor.

Primitive Accumulation and the Role of the Labor Surplus

"An increasing population," wrote Marx, is a "necessary condition" if "accumulation is to be a steady continuous process." This idea is subsequently qualified to mean growth of population "freed" from control over the means of production; that is, growth in the wage labor force and the industrial reserve army. The faster the growth in these aggregates, the more crises will appear as pauses within an overall trajectory of expansion (*Grundrisse,* 608, 764, 771; *Theories of Surplus Value,* pt. 2:47).

The key role of labor surpluses in Marx's thought cannot be overemphasized. Surpluses of labor must always be available if accumulation is to be sustained, while the processes of technological change which underlie crisis formation tend to produce surpluses of labor that cannot be absorbed. I concentrate for the moment on the prior process of production of the labor surplus. Where does this expansion in the exploitable population come from? In the "general law of capitalist accumulation," Marx divides the relative surplus population into three categories – latent, floating, and stagnant. The mobilization of latent reserves entails either primitive accumulation (the forcible separation of peasants, artisans, self-employed and independent producers, and even other capitalists from control over the means of production) or the substitution of family for individual labor (the employment of women and children in particular). A floating supply can be generated by any combination of sagging commodity production and labor-saving technological innovation. Taken in the context of natural population growth (itself not immune to the influence of capitalism's dynamic), these mechanisms must provide the fresh supplies of labor power to feed accumulation (the stagnant reserve army by virtue of its condition can play only a very small role).

Marx does not subject these processes to detailed scrutiny, but the flow of his logic points to certain conclusions. *Within* a particular civil society, viewed as a closed system, accumulation will accelerate until all latent elements are absorbed and the limits of natural population growth reached. Floating populations must then increasingly be relied upon as the source of an industrial reserve army. Society shifts from the trouble and turmoil of primitive accumulation and the destruction of precapitalist family relationships to the trauma of technologically induced unemployment. Both processes will be the focus of intense class struggles, though of a rather different sort. But resort to floating reserves is rather more problematic because it presupposes a strong dynamic of technological change, and this underlies, in the end, the tendency toward the overaccumulation of both capital and labor power. Though they may not be systematically aware of it, there is a distinct advantage to capitalists in exploiting latent rather than floating reserves. The more they depend on the latter, the deeper and more frequent crises will be.

To the degree that geographical expansion opens up access to fresh labor reserves, it can dampen the crisis tendencies of capitalism. Primitive accumulation on the exterior of civil society through the imposition of capitalist property relations by the state, the penetration of money relations and commodity exchange, and so forth, opens up new fields of action for overaccumulated capital. This was the main significance of the plantation economies, which capitalism brought under its sway (see above, sec. 2), and the new systems of colonial exploitation built up. Conversely, the labor surpluses so created could be imported from abroad. This, for Marx, was the significance of Ireland to English capitalism (*Ireland and the Irish Question*). Primitive accumulation in the former place furnished labor surpluses that could be used in the latter place to fuel accumulation, depress wages, and undermine the organized power of English workers (*Selected Correspondence*, 228–38). In the absence of slavery (or something akin to slavery, such as an indenture system), however, the importation of labor surpluses has to depend upon the free geographical mobility of labor power, and this implies the "abolition of all laws preventing the laborers from transferring from one sphere of production to another and from one local center of production to another" (*Capital* 3:196; Results of the Immediate Process of Production, 1013). But if the privilege of geographical mobility is conferred on the laborers from the exterior, it is hard to deny it to the floating reserves generated at home. The latter may emigrate to some free frontier or to wherever the money wage is highest. The significance of Marx's last chapter on colonization now strikes home with redoubled force. Primitive accumulation at the frontier is just as important as primitive accumulation at home or in some well-developed noncapitalist social formation abroad. Only in this way can the capitalist class as a whole ensure control over both the demand for and the supply of labor power on a global scale.

The "golden chain" that binds labor to capital can evidently be stretched a little in either direction, but only within strict limits. When labor is scarce the conditions of life of the laborer can improve somewhat, but at the price of triggering those technological adjustments that render capitalism more and more unstable and crisis-prone. In contrast, accelerating accumulation and the moderation of capitalism's internal contradictions depend upon the rapid expansion of the wage labor force through global primitive accumulation, the international mobilization of latent labor reserves, and other processes deeply disruptive of qualities of life, community, and standards of material well-being of the mass of the labor force. In either case, of course, the capital-labor relation remains intact – which is another way of saying that the geographical expansion of wage labor simply projects *the* fundamental contradiction of capitalism over ever-wider spheres and gives it greater and greater latitude of operation.

IV. THE GEOGRAPHY OF CAPITALIST ACCUMULATION: TOWARD A SYNTHESIS

The various programmatic sketches that Marx prepared before writing *Capital* invariably make reference to the geographical aspects of accumulation. In one such passage in the *Grundrisse,* for example, he begins with fundamental class relations, proceeds through topics such as town and country, the credit system, the "concentration of bourgeois society in the form of the state," and state debt, and concludes: "The colonies. Emigration. . . . The international relation of production. International division of labor. International exchange. Export and import. Rates of exchange. . . . The world market and crises" (108). Marx certainly thought that no theoretical account of capitalist development would be complete without integration of geographical aspects. And we have seen how he sporadically gestured in that direction. It remains, however, to evaluate the strengths and weaknesses of the ideas he did bequeath us and to move toward a more synthetic statement.

The implied synthesis can perhaps best be depicted as dual and intersecting contradictions that locate the underlying tensions that shape and propel the historical-geographical evolution of capitalism.

First Contradiction. Space can be overcome only through the production of space. We have encountered two rather different versions of this proposition. In the first we saw how the competitive thrust to diminish turnover time spills over into a compulsive pressure to break down spatial barriers and annihilate space with time. Close examination reveals how revolutions in the means of transportation and communications, how changes in the social division of labor and technology and in the locational requirements for capitalists and laborers, perpetually revolutionize space relations and the geographical landscapes of production, exchange, and consumption. Space cannot be overcome, however, without embedding productive forces in space. Tensions arise because the dead labor (capital) embedded in space ultimately becomes the barrier to be overcome. Capitalism has then no option but to destroy a part of itself in order to survive. The second version of this same proposition arises out of a consideration of foreign trade between communities (states) within the world market. The qualities of abstract labor come to dominate social life to the degree that different concrete labor processes in different communities are welded into a unity of global exchange. But concrete labor processes are perpetually modified in response to abstract requirements: "all old established national industries have been destroyed or are daily being destroyed," to be "dislodged by new industries whose introduction becomes a life and death question for all civilized nations." The circulation of capital in general and the creation of fresh productive forces within the community differentiate what simple exchange renders homogeneous. A permanent tension exists, therefore, between value as a representation of the universality of social labor and the particularities of concrete labor shaped as a competitive local response to the pressures exerted by competition in the world market. The homogenization of space through foreign trade is achieved, as it were, through processes that perpetually differentiate and transform the geographical division of labor.

Second Contradiction. The internal contradictions of capitalism can be resolved by a spatial fix, but in so doing capitalism transfers its contradictions to a wider sphere and gives them greater latitude. Overaccumulation of capital and labor power can be absorbed through geographical expansion, but this necessarily entails the reproduction of the social relations of capitalism in new geographical environments and spaces. Although capitalism can buy back time for itself out of the space it conquers, it cannot do so indefinitely nor avoid spreading the conditions for crisis formation over ever-wider spheres.

Intersection. The more fiercely capitalism is impelled to seek a spatial fix for its internal contradictions, the deeper becomes the tension of overcoming space through the production of space. The greater the overaccumulation and the faster the consequent pace of geographical expansion, the faster the pace of transformation of geographical landscapes. To the degree that those landscapes embody dead labor (capital) that has yet to be realized, so an increasing portion of that capital has to be destroyed in order to make way for new geographical configurations of production, exchange, and consumption. Capitalism requires increasing levels of self-destruction in order to survive. We here encounter the geographical dimension within Marx's basic proposition that "the universality towards which [capitalism] irresistibly strives encounters barriers in its own nature" (*Grundrisse,* 410).

Marx's stance with respect to the historical geography of capitalism has some distinctive virtues. The twin concepts of abstract and concrete labor for example, allow him to confront head-on a difficulty that has long bedeviled all forms of geographical enquiry: the relation between universal generalizations and the evident uniqueness of human activities and experience in particular places and times. A seemingly insuperable conceptual difficulty is thus transformed into a conceptual tension out of which theoretical elaborations can be evolved. Such a stratagem is intuitively appealing, since it explicitly recognizes how we are simultaneously in and of the world. But it also appears powerful enough to bear quite extraordinary conceptual fruits, chief of which, in my judgment, is a pathway toward the integration of geography and space relations into the general social theory that Marx proposed. That the proper harvesting of such a rich theoretical perspective has for so long been denied is, surely, one of the extraordinary and outstanding failures of an otherwise powerful Marxist tradition.

Our task, then, is to take these basic propositions concerning the contradictions inherent in the production of space under capitalist social relations and put them to work so as to understand how the historical geography of capitalism takes on its own particular and peculiar qualities. It is only in such a context that we can begin to reconstruct more detailed theoretical understandings of phenomena such as urbanization, regional development, and the geopolitics of uneven geographical development.

3

Class-Monopoly Rent, Finance Capital, and the Urban Revolution

In a stimulating and provocative work, Lefebvre (1970) argues that we ought to interpret the industrial revolution of the nineteenth century as a precursor of the "urban revolution" of the twentieth. He explains that by "urban revolution" he means: "the total ensemble of transformations which run throughout contemporary society and which serve to bring about the change from a period in which questions of economic growth and industrialization predominate to the period in which the urban problematic becomes decisive" (13).

Lefebvre is not explicit as to what is meant by "the ensemble of transformations," nor does he explain how and why capitalism is transformed so that questions of urbanization come to replace questions of economic growth and industrialization. Nor is Lefebvre very explicit when he argues that "the proportion of global surplus value formed and realized in industry declines" while the proportion realized "in speculation, construction and real estate development grows" (212). The thesis that this "secondary circuit of capital" is supplanting "the primary circuit of capital in production" is startling in its implications and obviously requires very careful consideration before being accepted or rejected. In this chapter, therefore, I shall attempt to shed some light on Lefebvre's hypotheses by examining how rent, and in particular class-monopoly rent, arises in the context of the urbanization process.

I. THE CONCEPT OF RENT IN AN URBANIZED WORLD

I take it as axiomatic that *value* arises out of those processes that convert naturally occurring materials and forces into objects and powers of utility to individuals in specific social and natural environments. In its simplest form, we can say that value arises out of production and is realized in consumption. But production and distribution cannot take place without (1) an elaborate social structure (encompassing the division of labor, the provision of socially necessary services, and so on), (2) a structure of social institutions through which individual and group activities can be coordinated, and (3) a certain minimum of physical infrastructure (communication links, utilities and the like). Any system of production and distribution requires, consequently, certain transfer payments to be made out of value produced to support socially necessary institutions, services, and physical infrastructures.

The history of the rental concept is strewn with arguments for and against the legitimacy of the transfer payment that rent represents (Keiper et al. 1961). In recent years, however, many appear to have been persuaded that rent is a kind of rationing device through which a scarce factor of production – land and its associated resources – is rationally and efficiently allocated to meet the productive needs of society (Wicksteed 1894). Rent is justified, according to this view, as a necessary coordinating device for the efficient production of value. The problem with this neoclassical argument is, however, that rent is regarded as a payment to a scarce "factor" (which is a "thing" concept) rather than as an actual payment to people. This reification may be convenient for purposes of analysis, but actual payments are made to real live people and not to pieces of land. Tenants are not easily convinced that the rent collector merely represents a scarce factor of production. The social consequences of rent are important and cannot be ignored simply because rent appears so innocently in the neoclassical doctrine of social harmony through competition.

There is a further point to be considered. In order for payments to be made, certain basic institutions are required. In our own society, private property arrangements are crucial; rent is, in effect, a transfer payment realized through the monopoly power over land and resources conferred by the institution of private property. Consequently, any examination of how rent originates and is realized cannot proceed without evaluating the performance of these supportive institutions.

But what is rent a payment for? The simplest answer is that it is a payment made by a user for the privilege of using a scarce productive resource that is owned by somebody else. But how does scarcity arise? Ever since production began to be organized systematically, human societies have recognized that many natural resources – understood as technical and cultural evaluations of nature – (Firey 1960; Spoehr 1956) are limited. There is a tendency, therefore, to think of scarcity as something inherent in nature, and on this basis we may be willing to concede that more should be charged for the use of productive fields and mines than for fields and mines of average productivity. On reflection, however, this conception of "natural wealth" and "scarcity" appears less satisfactory. There is little "natural wealth" that has not been prepared prior to production – the field has to be cleared and the mineshaft has to be dug. Relatively permanent improvements, such as the terracing of hillsides, the building up of soil fertility, and the draining of marshlands, may with time come to be regarded as "natural" resources for human use. In an urbanized world this problem becomes even more serious. Urbanization means the creation of relatively permanent resource systems (Harvey 1973, chap. 2). Human effort is, as it were, incorporated into the land as fixed and immobile capital assets that may last hundreds of years. Consequently, the high rent for a piece of land in the center of London may be due to its higher productivity, but that productivity has been *created* by the construction of the vast resource system that is London. Beause these relatively permanent fixed capital assets are highly localized in their distribution, the urbanization process has created scarcity where there was none before. If rent is a transfer payment to a scarce factor of production, then the urbanization process has also multiplied the opportunities for realizing rent.

The blurring of the distinction between natural and artificially created scarcity makes it difficult to distinguish between rent and profit. Are houses, for example, to be regarded as relatively permanent improvements incorporated into the value of the land, or are they better regarded as commodities commanding a profit on the capital outlay required to produce them? The answer to this question depends on what is meant by "relatively permanent." Housing has to be produced, and it has to be paid for as a commodity. Once this is done, however, the house may be regarded as a relatively permanent improvement incorporated into the value of the land. Buckingham Palace is a permanent improvement, whereas the suburban house just built is not yet in that happy state. It seems reasonable to think in similar fashion about other elements in the built form of the city – offices, shops, transport links, and so on.

The distinction between a mere transfer payment – rent – and profit on productive capital investment is difficult to keep in mind (see Chap. 4). The individual investor does not particularly care about the distinction; the overall rate of return on financial outlays is what matters. Money is put, therefore, where the rate of return is highest irrespective of whether productive activity is involved or not. If rates of return are high in the real estate and property markets, then investment will shift from the primary productive circuit of capital to this secondary circuit in a manner that would be consistent with Lefebvre's thesis. From the investor's point of view there is nothing to prevent such a shift. What has to be explained, however, is how returns can be higher on the secondary circuit over any length of time. For the fact that the distinction between productive and unproductive investment has disappeared from the investor's calculus does not negate the significance of such a distinction as a social fact. If all capital chases rent and no capital goes into production, then no value will be produced out of which the transfer payment that rent represents can come.

II. CLASS-MONOPOLY RENT, URBANIZATION, AND CLASS-MONOPOLY POWER

Rent can be charged for a variety of reasons. Marx's categories of differential, absolute and monopoly rent force us to consider the different kinds of situations out of which rent can arise (see Chap. 4). In this chapter I shall be concerned with what I call "class-monopoly rent."

This kind of rent arises because there exists a class of owners of "resource units" – the land and the relatively permanent improvements incorporated in it – who are willing to release the units under their command only if they receive a positive return above some arbitrary level (*Capital* 3: chap. 45). As a class these owners have the power always to achieve some minimum rate of return. The key concept here is *class power*. If landlords could not or would not behave in accordance with a well-defined class interest, then class-monopoly rents would not be realized. Landlords gain their class power in part from the fact that individually they can survive quite well without releasing all of the resource units under their command.

In nineteenth-century Europe, landlord power was essentially a residual from feudalism. Marx observed that it would be very much in the interest of the capitalist class to bring land and other productive resources under state ownership, since this would relieve the capitalist of the obligation of making any transfer payment to landed property (*Theories of Surplus Value,* pt. 2:44). It was unlikely, however, that capitalists would challenge the private property arrangements that allowed rent to be realized (and that provided the basis for the class power of landlords), since these arrangements also provided the necessary legal framework for entrepreneurial activity (Harvey 1982, chap. 11). But in an urbanized world, the distinction between capitalist and landlord has blurred concomitantly with the blurring of the distinctions between land and capital and rent and profit. We need, therefore, to adapt our categories to deal with the new complexities of extensive produced resource systems. But the same questions arise: are there owners of resource units (be they natural or artificial) who can and do behave so as to make it possible to realize rent? If so, what is the basis of their class power, how are they defining "class interest," and how are we to interpret their role in relation to the structure of social class in society as a whole?[1]

[1] It may be objected that I am using the concepts "class" and "class interest" far too freely and loosely. In what follows I shall use these concepts to refer to any group that has a clearly defined common interest in the struggle to command scarce resources in society. I shall use the phrase *social class* or *class structure* when referring to more general concepts of class in society. The notion of class-monopoly is made use of by Marx (*Capital* 3:194–95).

Landlords versus Low-Income Tenants

Suppose there exists a class of people who, by virtue of their income, social status, credit-worthiness, and eligibility for public assistance, are incapable of finding accommodation as homeowners or as residents in public housing. The existence of such a class is readily demonstrable in any large American or European city. This class of people has no alternative but to seek accommodation in the low-income rental market; they are trapped within a particular housing submarket. The needs of this class are provided for by a class of landlords. Landlordism varies, of course, from the old lady who rents an attic to the large-scale professional business operation. For purposes of exposition, let us assume that all rental accommodation is provided by a class of professional landlord-managers. This class has certain options as to where it puts its money, but much of its capital exists in the form of housing. On the basis of the potential yield of money on the capital market, professional landlords may set their expected rate of return on the estimated market value of their fixed capital assets at, say, 15 percent per annum. Suppose there is an abundance of low-income units in a particular city for some reason and that rates of return are in fact as low as 5 percent. A rational landlord strategy is to reduce maintenance, milk properties of value, and actively disinvest, using the money so extracted on the capital market where it earns, say, 15 percent. With declining maintenance, the housing deteriorates in quality, and eventually the worst units will be taken out of use – scarcity is successfully produced. Rents will gradually rise until the 15 percent rate of return is obtained (and there is nothing to stop rents going higher if circumstances allow). The class interest of the landlord is to obtain a minimum of 15 percent or else to find some way to get out of the market.

The class interests of landlord and tenant are clearly opposed to each other. If the quality of housing deteriorates and rents rise, tenants may seek accommodation elsewhere; but since they are, for the most part, trapped in this submarket, their power is limited in this respect. If they have some political power, they may seek to offset the class-monopoly power of landlords by imposing minimum housing standards or rent controls. If the effect of such legislation is to reduce landlord profits, landlords will respond by trying to transform the fixed capital (the house) into money to be used on the capital market. If prices are low, it will not be worthwhile to sell. Social, legal, and political pressures may make it difficult for the landlord to disinvest without severe social and fiscal penalties. Under these conditions, the landlord may well compromise and settle for a much lower rate of return. The tenants will then have achieved some kind of partial victory *vis-à-vis* the class-monopoly power of landlords. If, in contrast, tenants are politically weak, there is a shortage of suitable accommodation (because of in-migration or redevelopment); and if landlords can easily sell or transform to different uses (e.g., upper-income tenancies), then the landlord class will have very considerable power and will be able to raise their rate of return to well above 15 percent. With rising rents eating into an already limited disposable income, low-income tenants can respond only by subdividing space with the inevitable consequences – overcrowding and slum formation.

Class-interest conflict of this sort between tenants and landlords can be documented in any capitalist city (Chatterjee 1973; Sternlieb 1966; Milner-Holland Report 1965). The rate of return set through the working out of this conflict is best interpreted as a class-monopoly rent, even though the landlord usually thinks of it as a rate of return on capital investment. The realization of this rent depends upon the ability of one class-interest group to exercise its power over another class-interest group and thereby to assure for itself a certain minimum rate of return.

Speculator-Developers and Suburban Middle- and Upper-Income Groups

We now turn to a case that is rather more complex but that indicates that class-monopoly rents can be realized in all sectors of the housing market. Upper-income groups have a wide range of choice of housing as far as their income is concerned. But if their sense of social status and prestige is highly developed, then the producers of housing (who actively promote such thoughts on the part of the buyer) have an opportunity to realize a class-monopoly rent as these consumers vie with each other for prestigious housing in the "right" neighborhoods (see Chap. 5). Middle-income groups may have less choice. In many American cities, for example, they have moved to suburbia, in part because they were hooked on the suburban dream, but also because social changes in the city – the influx of a low-income "lumpenproletariat," the decline of city services, falling property values, the withdrawal of financial support for whole neighborhoods, and declining employment opportunities – have given them a hefty push by a process that I have elsewhere dubbed "blow-out" (Harvey 1973, chap. 5).

The realization of a class-monopoly rent depends, however, on the existence of a class of speculator-developers who have the power to capture it.[2] In a free market economy, speculator-developers perform a positive service. They promote an optimal timing of land-use change, ensure that the current

[2] The term *speculator-developer* is here used generically to refer to all those individuals and institutions that operate in the land and property markets with a view to realizing gains through ultimate sale or change in land use. In practice there may be considerable division of labor in this activity, while different institutions operate under different constraints. See, for example, the difference between entrepreneurs and the relics of the feudal order – the crown, the church, etc. – described in Counter-Information Services (1973).

value of land and housing reflects expected future returns, seek to organize externalities to enhance the value of their existing developments, and generally perform a coordinating and stabilizing function in the face of considerable market uncertainty (Neutze 1968; Hall et al. 1973). The role of speculator-developer is, in fact, integral and essential to the workings of a capitalist economy. Since the urbanization process relates to economic growth in general, the speculator-developer, who is, in effect, the promoter of urbanization, plays a vital role in promoting economic growth. Certain institutional supports are necessary, however, if this role is to be performed effectively. The exact nature of these supports will vary from country to country but they must do two things: (1) they must reduce the uncertainty in land-use competition, usually through some form of governmental regulation – planning or zoning controls, provision of infrastructure, etc.; and (2) they must encourage wealthy groups – those who can afford to wait for land to "ripen" – to participate as speculator-developers, usually by offering convenient and advantageous tax arrangements. The first support permits speculator-developers to form reasonable expectations about the future, while the second ensures that only people with sufficient resources undertake the task of coordinating and stabilizing land-use change.

Class-monopoly rents can be realized by speculator-developers only if they possess mechanisms for expressing their collective class interest. The necessary institutional supports in fact provide these mechanisms. In the United States, for example, speculator-developers usually realize monopoly-rents through the manipulation of zoning decisions. Political control of suburban jurisdictions by speculator-developers is quite general in the United States; as Gaffney (1973) notes, suburban jurisdictions provide one of the most effective of all cartel arrangements with respect to land-use decisions. Political corruption also plays a role which, in a market economy, can be viewed positively, since it frequently loosens up the supply of land from the excessive rigidities of land-use regulation by bureaucratic fiat. Without a certain minimum of governmental regulation and institutional support, however, the speculator-developer could not perform the vital function of promoter, coordinator and stabilizer of land-use change. Without such an interest group to perform these functions, suburban development would degenerate into chaos and finance capital would be forced to withdraw investment from the suburbanization process. The effect of such a withdrawal upon economic growth in general, effective demand in general, and the capitalist market system as a whole would, of course, be catastrophic.

The level of class-monopoly rent realized by speculator-developers depends upon the outcome of the conflict of interest between them and the various consumer groups who confront them in the market. If the speculator-developer can persuade upper-income groups of the virtues of a certain kind of housing in a particular neighborhood, gain complete control over the political process, and so on, then the advantage lies with the speculator-developer. If consumers are unimpressed by the blandishments of the speculator-developers and have firm control over the political mechanisms for land-use regulation and the provision of infrastructure, then the class-monopoly power of the speculator-developers will be contained. But if certain minimum rates of return are not realized, the speculator-developer will pull out of the business until rates of return rise. What the minimum must be is difficult to say – but in the United States a 40 percent rate of return is not regarded as abnormal.

The two cases we have examined – the landlord versus the low-income tenant and the speculator-developer versus the middle- and upper-income consumer – provide us with certain insights into the meaning of class-monopoly rent and class-monopoly power in the context of urbanization. First, this form of rent appears inevitable in capitalistically organized land and housing markets. Second, the transfer payments that result from class-monopoly rents are structured in certain important respects. Suppose the landlord lives in suburbia and as a resident there gives up a class-monopoly rent to the speculator-developer. Notice that the rent realized from a low-income tenant has been passed on, in this example, to the speculator-developer via the landlord. It is unlikely, bordering on the impossible, for rent realized by the speculator-developer to be passed on to the low-income tenant. It seems reasonable to postulate, therefore, a hierarchical structure of some sort through which class-monopoly rents percolate upward but not downward. At the top of this hierarchy sit the financial institutions. And so the question arises, how does this hierarchy arise and what is its *raison d'être?*

III. THE HIERARCHICAL INSTITUTIONAL FRAMEWORK FOR COORDINATING ACTIVITIES IN THE HOUSING MARKET

I shall begin by stating a general proposition: The hierarchical institutional structure through which class-monopoly rents are realized is a necessity if housing market activity is to be coordinated in a way that helps to avoid economic crisis. The problem with seeking to validate this proposition is that institutional arrangements vary markedly from country to country. But all capitalist economies must, of necessity, possess elaborate devices to integrate national and local aspects of economies, to integrate individual decisions with the needs of society as a whole. Any society must possess, in short, formalized human practices that resolve the social aggregation problem (Harvey and Chatterjee 1974). These formalized human practices are manifest in a structure of financial and governmental institutions which, I shall argue, create the basis for class-monopoly power in the land and property markets. To explore this proposition I shall examine institutional structures in the United States and consider how these affect events in Baltimore in particular.

National institutions of government and finance do not operate without a purpose; they seek, by and large, to ensure the reproduction of society and to deal with any problems that may arise in an orderly and nondisruptive manner. In a capitalist society this means a policy directed toward the orderly accumulation of capital, economic growth, and the reproduction of the basic social and political relationships of a capitalist society. In the housing market these general concerns are translated into three typical concerns for national housing policy:

1. To ensure orderly relationships between construction, economic growth, and new household formation
2. To ensure short-run stability and iron out cyclical swings in the economy at large by using the construction industry and the housing sector as a partial Keynesian regulator
3. To ensure domestic peace and tranquillity by managing the distribution of welfare in society through the provision of housing

In the United States these concerns have been embedded in policy goals which have, by and large, been successfully met since 1930. Economic growth has been accompanied and to some degree accomplished by rapid suburbanization – a process that has been facilitated by national housing policies conducted through the Federal Housing Administration (FHA). Much of the growth in GNP, both absolute and per capita, since the 1930s has been wrapped up in the suburbanization process (taking into account the construction of highways and utilities, housing, the effective demand for automobiles, gasoline, and so on). Cyclical swings in the economy have been broadly contained since the 1930s, and the construction industry appears to have functioned effectively as a major countercyclical tool. The evident social discontent of the 1930s has largely been defused by a government policy that has created a large wedge of middle-income people who are now "debt-encumbered homeowners" and consequently unlikely to rock the boat. The discontent of the 1960s exhibited by the blacks and the urban poor provoked a similar political response in the housing sector – a response that has not provided a "decent house in a decent living environment" (as congressional legislation perennially puts it) for many of the poor but that has successfully created a debt-encumbered class of black homeowners; the social instability of the 1960s certainly appears to have been held in check. It appears, then, that national policies are designed to maintain the existing structure of society intact in its basic configurations while facilitating economic growth and capitalist accumulation, eliminating cyclical influences, and controlling social discontent.

How are these national policies transmitted to the locality, and how do individuals come to incorporate them into their decisions? Federal, state and local governments form a three-tiered political hierarchy, and an independent bureaucracy is attached to each level. The federal bureaucracy is itself hierarchically organized, however, so that it is in a position to relate national policies to local housing markets. The Federal Housing Administration administers a wide range of government programs and operates autonomously from bureaucracies created at the state and local levels. But in the United States the main mechanism for coordinating national and local, individual, and societal activities lies in the hierarchical structure of financial institutions operating under governmental regulation. This structure is exceedingly complex, and I shall not attempt to detail it here. It is important to note one feature of it, however. Certain kinds of institution – the state and federally chartered savings and loan institutions – operate solely in the housing sector. They were initially designed to "promote the thrift of people locally to finance their own homes and the homes of their neighbors." Some of these institutions are community-based and depositor-controlled and operate on a nonprofit basis. They are, of course, affected by money market conditions and government regulation. These institutions contrast with the mortgage banks, savings banks, and commercial banks, which are oriented to profits or to the expansion of their business. All of these institutions, however, operate together to relate national policies to local and individual decisions and, in the process, create localized structures within which class-monopoly rents can be realized.

The Baltimore situation during the 1960s and early 1970s demonstrates the point. As of 1970, the metropolitan area had a population of approximately two million; 900,000 in Baltimore City and 600,000 in the largest suburban jurisdiction – Baltimore County – which surrounds Baltimore City on almost all sides. The political machine in Baltimore County has been dominated by speculator-developer interests who have, until recently, been able to manipulate the zoning laws more or less at will in order to realize speculative gains. Political corruption is usual (Spiro Agnew was once county executive). All that is necessary for the realization of class-monopoly rents is some generally sustained demand for new housing (through population increase or new household formation). There is a further point to be considered, however. The investment climate during the 1960s and early 1970s was radically different in Baltimore County and Baltimore City. All of the institutions looked collectively on the former as an area of growth and expansion compared to the city, which was looked upon as an area that was at best stable and at worst in the process of rapid decline. The consequent channeling of investment funds to the county and the general reluctance to invest in the city turned out to be a self-fulfilling prediction to which middle-income groups were forced to respond by migrating from the city to the county, where the speculator-developer eagerly awaited them. In this fashion the conflict between city and suburb in the United States contributed to the realization of class-monopoly rents on the suburban fringe.

But there was also a geographical structure to the housing market in Baltimore City which further contributed to the potential for realizing class-monopoly rents. This geographical structure was produced by the interacting policies of financial and governmental institutions. To demonstrate this point, Baltimore City is divided into thirteen submarkets that can be further aggregated into eight submarket types (fig. 10). Data for 1970 concerning the financing of housing in each of these submarkets, together with some socioeconomic information, are presented in tables 1 and 2. The housing market in Baltimore City appears highly structured geographically with respect to the type of institutional involvement as well as with respect to the insurance of home purchases by the Federal Housing Administration (fig. 11). Let us consider the main features of this structure.

1. The *inner city* was dominated by cash and private loan transactions with scarcely a vestige of institutional or governmental involvement in the used housing market. This submarket was the locus of that conflict between landlord and low-income tenant to which we have already alluded. There was at any time a surplus of housing in this submarket which led to active disinvestment (there were several thousand vacant structures in this submarket). Professional landlords were anxious to disinvest, but they still managed to get a rate of return of around 13 percent (Chatterjee 1973). The tenants were low-income and for the most part black. They were poorly organized, exercised little political control, and were effectively trapped in this submarket. Class-monopoly rents were here realized by professional landlords who calculated their rate of return to match the opportunity cost of their capital.

2. The white *ethnic areas* were dominated by homeownership, financed mainly by small, community-based savings and loan institutions that operate without a strong profit orientation and which really do offer a community service. As a consequence little class-monopoly rent was realized in this submarket, and reasonably good housing was obtained at a fairly low purchase price, considering the fairly low incomes of the residents.

3. The black residential area of *West Baltimore* was essentially a creation of the 1960s. Low- to moderate-income blacks did not possess local savings and loan associations, were regarded with suspicion by all other financial institutions,

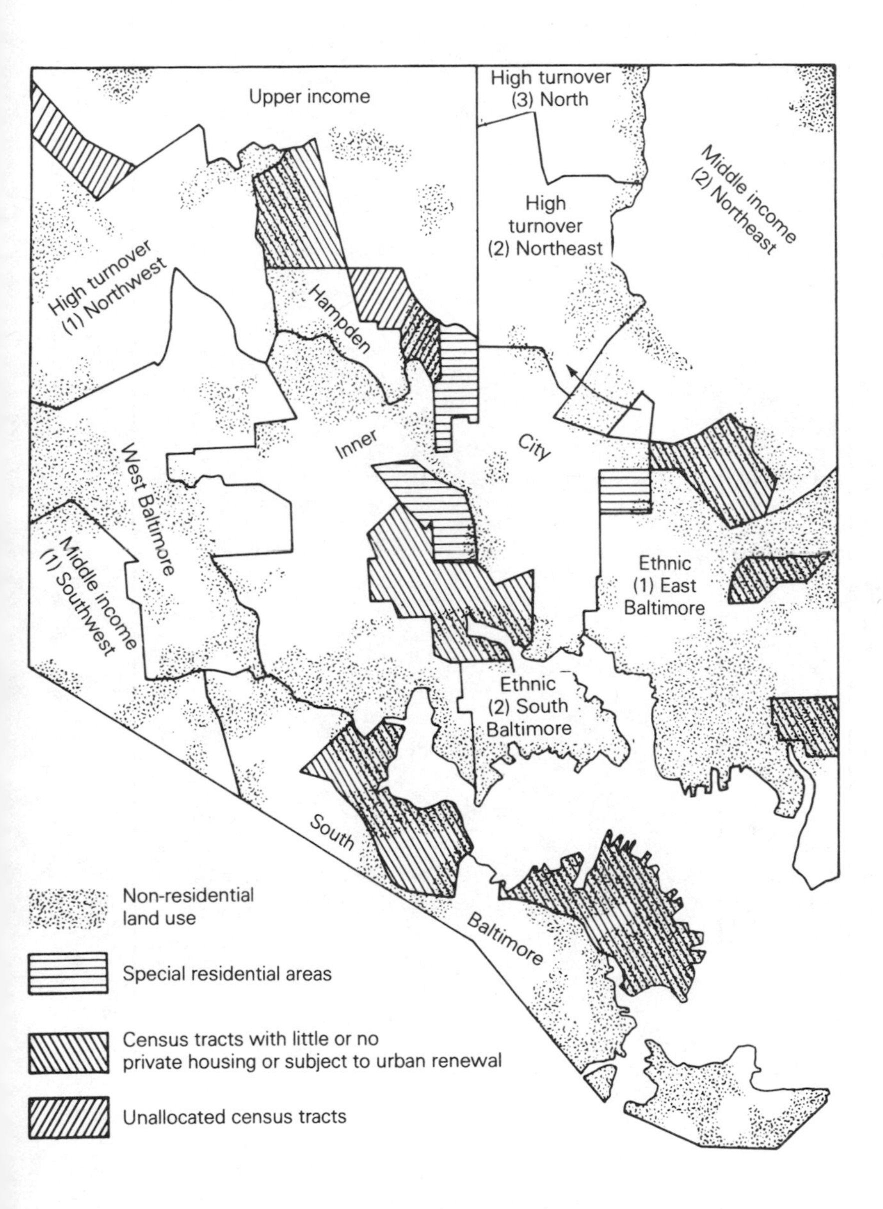

Fig. 10. The housing submarkets of Baltimore City, 1970

Table 1. Housing Submarkets in Baltimore City, 1970 (Institutional Activity)

Sectors	Total Houses Sold	Sales per 100 Properties	% Transactions by Source of Funds: Cash	Pvt.	Federal S. & L.	State S. & L.	Mortgage Bank	Community Bank	Savings Bank	Other[a]	% Sales Insured FHA	VA	Average Sale Price($)[b]
Inner city	1,199	1.86	65.7	15.0	3.0	12.0	2.2	0.5	0.2	1.7	2.9	1.1	3,498
1. East	646	2.33	64.7	15.0	2.2	14.3	2.2	0.5	0.1	1.2	3.4	1.4	3,437
2. West	553	1.51	67.0	15.1	4.0	9.2	2.3	0.4	0.4	2.2	2.3	0.6	3,568
Ethnic	760	3.34	39.9	5.5	6.1	43.2	2.0	0.8	0.9	2.2	2.6	0.7	6,372
1. East Baltimore	579	3.40	39.7	4.8	5.5	43.7	2.4	1.0	1.2	2.2	3.2	0.7	6,769
2. South Baltimore	181	3.20	40.3	7.7	7.7	41.4	0.6			2.2	0.6	0.6	5,102
Hampden	99	2.40	40.4	8.1	18.2	26.3	4.0		3.0		14.1	2.0	7,059
West Baltimore	497	2.32	30.6	12.5	12.1	11.7	22.3	1.6	3.1	6.0	25.8	4.2	8,664
South Baltimore	322	3.16	28.3	7.4	22.7	13.4	13.4	1.9	4.0	9.0	22.7	10.6	8,751
High turnover	2,072	5.28	19.1	6.1	13.6	14.9	32.8	1.2	5.7	6.2	38.2	9.5	9,992
1. Northwest	1,071	5.42	20.0	7.2	9.7	13.8	40.9	1.1	2.9	4.5	46.8	7.4	9,312
2. Northeast	693	5.07	20.6	6.4	14.4	16.5	29.0	1.4	5.6	5.9	34.5	10.2	9,779
3. North	308	5.35	12.7	1.4	25.3	18.1	13.3	0.7	15.9	12.7	31.5	15.5	12,330
Middle income	1,077	3.15	20.8	4.4	29.8	17.0	8.6	1.9	8.7	9.0	17.7	11.1	12,760
1. Southwest	212	3.46	17.0	6.6	29.2	8.5	15.1	1.0	10.8	11.7	30.2	17.0	12,848
2. Northeast	865	3.09	21.7	3.8	30.0	19.2	7.0	2.0	8.2	8.2	14.7	9.7	12,751
Upper income	361	3.84	19.4	6.9	23.5	10.5	8.6	7.2	21.1	2.8	11.9	3.6	27,413

Source: City Planning Department Tabulations from Lusk Reports.

[a] Assumed mortgages and subject to mortgage.

[b] Ground rent is sometimes included in the sale price, and this distorts the averages in certain respects. The relative differentials between the submarkets are of the right order, however.

Table 2. Housing Submarkets in Baltimore City, 1970 *(Census Data)*

	Median Income[a]	*% Black-occupied D.U.'s*	*% Units Owner Occupied*	*Mean $ Value of Owner Occupied*	*% Renter Occupied*	*Mean Monthly Rent*
Inner city	6,259	72.2	28.5	6,259	71.5	77.5
1. East	6,201	65.1	29.3	6,380	70.7	75.2
2. West	6,297	76.9	27.9	6,963	72.1	78.9
Ethnic	8,822	1.0	66.0	8,005	34.0	76.8
1. East Baltimore	8,836	1.2	66.3	8,368	33.7	78.7
2. South Baltimore	8,785	0.2	64.7	6,504	35.3	69.6
Hampden	8,730	0.3	58.8	7,860	41.2	76.8
West Baltimore	9,566	84.1	50.0	13,842	50.0	103.7
South Baltimore	8,941	0.1	56.9	9,741	43.1	82.0
High turnover	10,413	34.3	53.5	11,886	46.5	113.8
1. Northwest	9,483	55.4	49.3	11,867	50.7	110.6
2. Northeast	10,753	30.4	58.5	11,533	41.5	111.5
3. North	11,510	1.3	49.0	12,726	51.0	125.1
Middle income	10,639	2.8	62.6	13,221	37.5	104.1
1. Southwest	10,655	4.4	48.8	13,470	51.2	108.1
2. Northeast	10,634	2.3	66.2	13,174	33.8	103.0
Upper income	17,577	1.7	50.8	27,097	49.2	141.4

Source: 1970 census.
[a] Weighted average of median incomes for census tracts in submarket.

and in the early 1960s were discriminated against by the FHA. The only way in which this group could become homeowners was by way of something called a "land-installment contract," which works as follows. A speculator purchases a house for, say, $7,000; adds a purchase and sales commission, various financing charges, and overhead costs; renovates and redecorates the property; and finally adds a gross profit margin of, say, 20 percent. The house is then sold for, say, $13,000. To finance the transaction, the speculator interposes his credit rating between that of the purchaser and the financial institutions, takes out a conventional mortgage up to the appraised value of the house (say, $9,000), borrows another $4,000, and then packages a $13,000 loan for the buyer. The speculator retains title to the property to secure the risk but permits the "buyer" immediate possession. The monthly payments cover the interest charges on the $13,000 plus the administrative charges, and a small part is put to redeeming the principal. When the purchaser has redeemed $4,000 (after, say, ten or fifteen years), a conven-

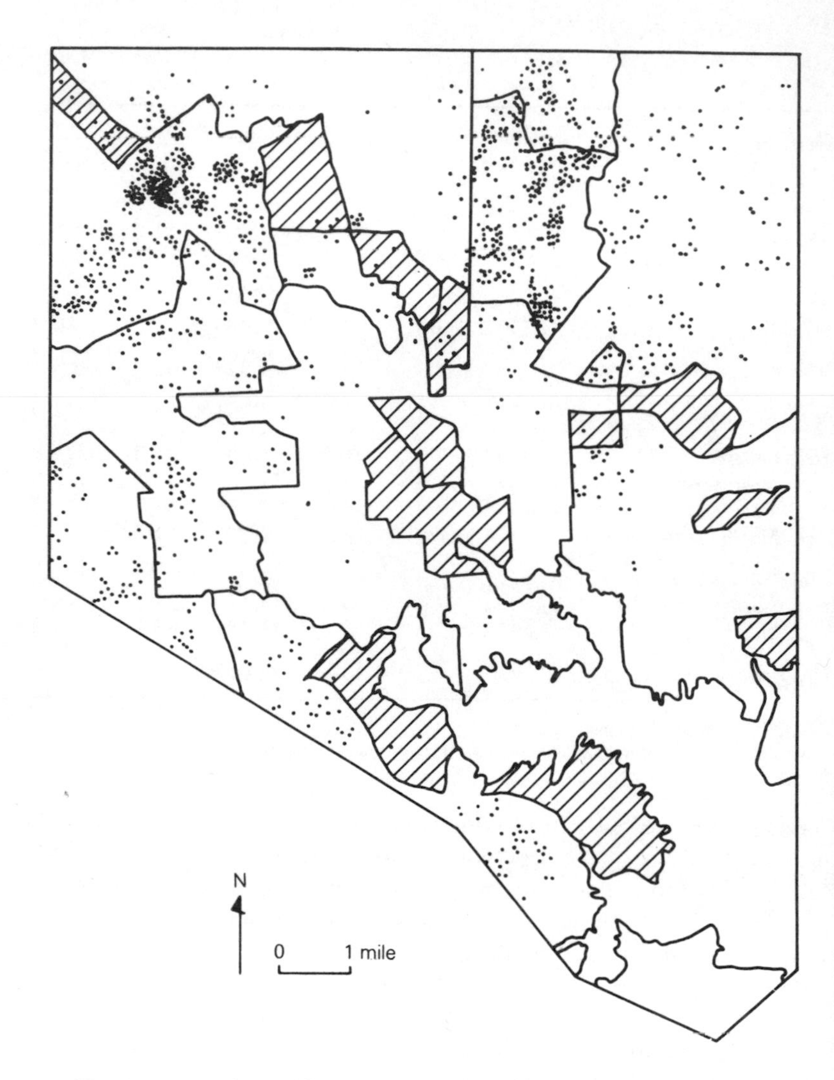

Fig. 11. Distribution of FHA-insured mortgages across housing submarkets in Baltimore City, 1970.
(Source: Lusk Reports.*)*

tional mortgage at the appraised value of $9,000 may be obtained. At that juncture the purchaser will get title and can start to build equity in the house (see Grigsby et al. 1971, chap. 6).

This procedure is perfectly legal, and it was in effect the only way in which low- or moderate-income blacks could become homeowners in the early 1960s. There were many transactions of this sort in West Baltimore. The problem was that a comparable house sold to a person in a comparable income bracket in white ethnic areas cost $7,000 compared to the $13,000 registered in the black community. Blacks consequently regarded themselves as exploited and paying the "black tax," which was nothing more nor less than class-monopoly rent realized by speculators as they took advantage of a particular mix of financial and governmental policies compounded by problems of racial discrimination. But a new submarket was formed in West Baltimore by means of the land-installment contract; and in the process strong pressures were exerted on white middle-class groups to move to suburbia where the speculator-developers waited, all too willing and able to accommodate them.

The political conflict over the use of the land-installment contract in Baltimore came to a head in the late 1960s. In the process the black communities learned that the speculator was creeping in where financial and governmental institutions refused to tread and that the problems of speculation could not be divorced from the activities of the financial and governmental institutions. Since the 1960s the land-installment contract has declined as a form of financing. But the speculator has not disappeared from the scene; rather, he now has other instrumentalities at his disposal.

4. The areas of *high turnover* were serviced mainly by a combination of mortgage banker finance and FHA insurance, which was doing in 1970 what the land-installment contract did in the 1960s. Various programs were initiated in the late 1960s to try and create a debt-encumbered, socially stable class of homeowners amongst the black and the urban poor. These programs, together with administrative directives to end discriminatory practices against blacks, led to the creation of an FHA-insured, mainly black, fairly low-income housing submarket. The main tool in Baltimore was the FHA 221 (d)(2) program (D2s), which permitted the financing of homeownership for low- or moderate-income persons who had no money for a down payment. FHA insurance in Baltimore in 1970 was, for the most part, of the D2 variety (see fig. 11).

In the high turnover submarkets created by these programs there were plenty of opportunities for the speculator to realize a class-monopoly rent. Operating through the D2 programs made it less easy to extract the "black tax"; but if whites moved (as they were likey to do if a low-income black family moved in), speculators could pick up houses at less than the appraised value, put in some cosmetic repairs to meet FHA quality standards, and sell through the D2 program. If FHA quality control standards were poor (or if speculators could corrupt the administration of them), then class-monopoly rents could be realized as a white "exit tax" and a black or low-income "entry tax." In some cities, such as Detroit, New York, and Philadelphia, the windfall profits to speculators were enormous, largely through the corruption of FHA programs (Boyer 1973). In Baltimore, the submarket created by the land-installment contract in the 1960s was extended in areas of high turnover by speculator activity in conjunction with mortgage banker finance and the FHA D2 programs.

5. The *middle-income* submarkets of Northeast and Southwest Baltimore were typically the creation of the FHA programs of the 1930s. By the 1960s homeownership was being financed conventionally by federal savings and loan institutions and by some of the smaller ethnic savings and loan institutions which helped to finance migration from the older ethnic areas of the city to the newer housing of Northeast Baltimore. The inner edge of this submarket was under some pressure, however, and financial institutions were extremely sensitive about risks in these areas. As a consequence they tended to withdraw support from an area if they perceived it to be threatened in any way. By doing so they created a vacuum in housing finance into which the speculator moved, backed by the FHA programs and mortgage banker finance. There was a good deal of political friction in these boundary zones and a political struggle to preserve the middle-income submarkets from erosion at the edges – an erosion that inevitably led middle-income groups to search for housing opportunities in the suburbs.

6. The more affluent groups made use of savings and commercial banks to a much greater degree and rarely resorted to FHA guarantees. Such groups usually had the political and economic power to fend off speculative incursions and it was unlikely that they would move except as the result of their own changing preferences or from declining services. Class-monopoly rents were realized in this submarket largely because of prestige and status considerations.

This geographical structure of submarkets in Baltimore around 1970 formed a decision environment in the context of which individual households made housing choices. These choices were likely, by and large, to conform to the structure and to reinforce it. The structure itself was a product of history. In the long run we find that the geographic structure of the city is continuously being transformed by conflicts and struggles generated by the ebb and flow of market forces; the operations of speculators, landlords, and developers; the changing policies of governmental and financial institutions; changing tastes; and the like. But in the short run the geographic structure is rather fixed, and it is this rigidity that permits class-monopoly rents to be realized within submarkets (as classes of providers face classes of consumers) and between submarkets as a variety of processes seek to erode the boundaries of the submarkets themselves (every submarket has its speculator-developer fringe). In some parts of the city these conflicts may be dormant at times – boundaries may be stabilized (often with the help of natural or artificial barriers), and accommodation between opposing forces may be reached within submarkets. But it would be rare indeed to find a city in which no such conflicts were occurring.

IV. CLASS-MONOPOLY RENT, ABSOLUTE SPACE, AND URBAN STRUCTURE

Class-monopoly rents arise because the owners of resource units have the power always to exact a positive return. Ricardo thought that *absolute* rent could exist only on an island where all resource units were employed and on which there was an absolute scarcity. The Baltimore materials indicate that the man-made resource system created by urbanization is, in effect, a series of man-made islands on which class monopolies produce absolute scarcities. Absolute spaces, created by human practices, are essential, it seems, to the realization of class-monopoly rent. Absolute spaces can be constructed by dividing space into parcels and segments, each of which can then be regarded as a "thing-in-itself" independent of other things (Harvey 1973, chap. 5). The private property relation is, of course, the most basic institution by means of which absolute spaces are formally created. Political jurisdictions define collective absolute spaces, which may then be carved up by the bureaucratic regulation of land use. All of these forms of absolute space create the possibility of realizing class-monopoly rents. But it is primarily through the informally structured absolute spaces of submarkets that such rents are realized.

The implications of this for residential structure are of interest. Residential differentiation in urban areas has long been explained in terms of social ecological processes, consumer preferences, utility-maximizing behaviors on the part of individuals, and the like. The Baltimore evidence suggests that financial and governmental institutions play an active role in shaping residential differentiation and that the active agent in the process is an investor seeking to realize a class-monopoly rent from the circulation of revenues (see Chap. 4). The rèlationship between traditional explanations of urban residential differentiation and this interpretation is complex. The small neighborhood savings and loan institution in Baltimore, for example, is in effect a community institution that fits neatly into a social ecological view of urban community structure. But most housing finance comes from institutions seeking profits or the expansion of business. Faced with a choice between supporting a risk-absorbing landlord operation and a vulnerable homeowner in the inner city, business rationality dictates support of the former at the expense of the latter. Not all financial institutions exhibit a totally cold market rationality – they will grant personal favors (usually to people of the same social class, however), and they will sometimes actively support neighborhoods (often to procure a desirable stability in a particular submarket). But the options of the profit-maximizing or expansion-conscious financial institution are limited. The hidden hand of the market, and in particular the prospects for realizing class-monopoly rents, will inexorably guide them in certain directions. And as a result these institutions become a fundamental force in shaping the residential structure of the city.

This is not to say that considerations of race and ethnicity, social status and prestige, life-style aspirations, community and neighborhood solidarity, and the like are irrelevant to understanding residential differentiation. Ironically, all of these features *increase* the potential for realizing class-monopoly rent because they help to maintain the islandlike structure, to create the absolute space of the parochially-minded community. Indeed, a strong case can be made for regarding consumer preferences as being produced systematically rather than as arising spontaneously (as neoclassical economic doctrines appear to envisage via the myth of consumer sovereignty). The simplest manifestation of this is the use of techniques of persuasion to convince upper-income people of the virtues of living in a "smart" house in the "right" neighborhood (see Chap. 5). But there is also a deeper process at work. Financial institutions and government manage the urbanization process to achieve economic growth and economic stability and to defuse social discontent (see sec. 3). If these aims are to be realized, then new modes of consumption and new social wants and needs will have to be produced whether people like it or not. If these new modes of consumption and new social wants and needs do not arise spontaneously, in a manner that fits with the overall necessities of capitalist society, then people will have to be forced or cajoled to accept them. The urbanization process achieves this end quite successfully. By structuring and restructuring the choices open to people, by creating distinctive decision environments, the urbanization process forces new kinds of choice independently of spontaneously arising predilections (cf. Chap. 8).

If the physical dynamic of urbanization is powered by financial and governmental institutions, mediated by speculator-developers and speculator-landlords in pursuit of class-monopoly rent, and necessitated by the overriding requirement to reproduce the capitalist order, then it may not be too fanciful to suggest that distinctive "consumption classes," "distributive groupings," or even "housing classes" may be produced at the same time (see

Giddens 1973; Rex and Moore 1975). Individuals can, of course, strive or choose to join one or another "distributive grouping," or shift (if they can) from one "consumption class" to another. In like manner they can strive or choose (depending on their circumstances) to move from one housing submarket to another. What individuals cannot choose, however, is the structure of the distributive grouping or the structure of the housing submarkets – these are dictated by forces far removed from the realms of consumer sovereignty. The general proposition we are here led to is an intriguing one: in producing new modes of consumption and new social wants and needs, the urbanization process concomitantly produces new distributive groupings or consumption classes, which may crystallize into distinctive communities within the overall urban structure. This topic will be taken up again in section 5.

The Baltimore materials suggest another startling conclusion. The class-monopoly rent gained in one submarket is not independent of its realization elsewhere, and certain strong multiplier effects can be detected. Suppose, for example, that there is a speculative boom in the inner city through which new submarkets are formed out of existing neighborhoods and that the old residents of these neighborhoods are forced to seek housing opportunities in suburbia. Then, the greater the class-monopoly rent earned by the inner-city speculator, the greater the opportunity to realize rent on the suburban fringe. Multiplier effects of this sort may be captured by the same financial institutions or, in some cases, by the same entrepreneur. If there is no conscious collusion to generate the multiplier effect, the calculus of profits and losses, of expectations and perceived risks, will function as a hidden-hand regulator to achieve the same results.

These conclusions are, of course, geographically and institutionally specific to Baltimore and the United States. But a cursory examination of the literature suggests that they may be generalized to all advanced capitalist nations.[3] Whether or not this is the case must be proved by future research. It seems likely, however, that the processes are general but that the manifestations are particular because the institutional, geographical, cultural, and historical situations vary a great deal from place to place. In other words, the processes are general, but the circumstances are unique to each case and so, consequently, are the results.

If the multiplier effects of the realization of class-monopoly rents are

[3] For comparable materials on London see, for example, Hall et al. (1973), The Milner-Holland Report (1965), Pahl (1975), Counter-Information Services (1973), and Marriott (1967). The point here is, of course, that the large number of vacant houses in the center of Baltimore is a vivid contrast with the situation in London; but the process of gentrification in London is as much a manifestation of the process of realizing class-monopoly rent as is the land-installment contract and speculation with the D2s in Baltimore.

general, then we have a partial explanation of how investment can shift continuously over time from the primary to the secondary circuit of capital as Lefebvre hypothesizes. Governmental and financial institutions are forced to operate in certain ways if individual behaviors are to be coordinated and integrated with national and societal requirements. Urbanization, itself a product of these requirements, creates islands of opportunities for realizing class-monopoly rents. And the quest for this rent generates a multiplier effect that makes it even more profitable in the short run to shift investment into the land, housing, and property markets. Such a shift helps to explain the industrial stagnation, particularly evident since the late 1960s, in the advanced capitalist nations as investment shifts from the production of value to the attempt to realize it. In the short run such a shift is possible because it is possible to milk value produced in past periods for purposes of current realization (which means, however, a continuous decay in the quality of urban environments). But in the long run such a shift is doomed to failure, for if value is not produced, then how can it possibly be realized?

V. CLASS SYSTEM, CLASS STRUCTURE, AND CLASS INTEREST IN THE POLITICAL ECONOMY OF URBANISM

We shall now consider the relationship between the concept of class interest as it arises in the context of urbanization (and as it is used in this chapter) and more general concepts of class structure and class antagonism. It is useful to distinguish at the outset between the concept of *subjective* class, which describes the consciousness different groups have of their position within a social structure, and the concept of *objective* class, which, in Marx's schema, describes a basic division within capitalism between a class of producers and a class of appropriators of surplus value (for a recent discussion of this topic, see Giddens 1973). The former class includes both productive labor and the labor that is socially necessary but unproductive (for example, labor contributing to circulation, realization, administration, and the provision of socially necessary services). The meaning attached to class interest in this chapter stems from the fact of certain conflicts around the realization of class-monopoly rent. I am, therefore, working at the level of subjective class interest, and the task is to relate these diverse class interests to the concept of objective class.

Traditionally, rent is viewed as a transfer payment from capitalist producers to a rentier class that gains its power as a residual of feudalism. But we are here concerned with rent extracted from the community out of the consumption process rather than out of the production process. It relates more to the circulation of revenues than it does to the circulation of capital (see Chap. 4). This extraction generates a species of community conflict which has become widespread with the progress of urbanization in advanced capitalist countries. This kind of conflict contrasts, superficially at least, with the more traditional work-based conflicts over the immediate production of value. We can observe as a consequence some curious dichotomies. Community-based organizations rarely offer support in a work-based conflict (such as a strike), and work-based organizations (such as the trade unions) rarely offer active support to community groups in conflict over, for example, the realization of class-monopoly rent. Individuals may in fact switch roles with respect to such conflicts – a work-based radical may be a community-based conservative (and vice versa). The place of work also tends to be male-dominated space compared with the female-dominated residential space. Sex roles may become intertwined as a male work-based radical acts conservatively with respect to a female community-based radical. Such conflicts may be internalized within the family. The geography of human activity within large metropolitan areas appears to generate curious transformations and inversions to create a complex geography of subjective class-consciousness. The expression of class interest around community issues cannot, therefore, be interpreted as a simple reflection of class interest at the point of production.

Yet class interest can be equally strong and express analogous goals at the point of production and within the community. Workers may seek worker control, while residents may seek community control. Both goals express a basic felt need on the part of individuals to control the social conditions of their existence. But under urbanization the two goals become divorced from each other. A far more cohesive basis for political power exists when community-based and work-based interests coincide (for example, in mining communities and in other situations characterized by *industrial* rather than *advanced urban* forms of social organization). Marx thought that large concentrations of population would heighten class-awareness. But under urbanization class-consciousness appears to have become fragmented (see *Consciousness and the Urban Experience*).

Community-based class interest always tends to be parochialist in its perspectives. The community is regarded as a "thing-in-itself" independent of other things – it is regarded as an absolute space, as something to be preserved and defended against external threat. From such a standpoint flows a form of community conflict which is essentially internecine – it pits community against community so that the average condition of communities is not altered one whit. What one community gains another loses. The sequence of wins and losses merely serves to perpetuate the defensiveness and competitiveness of the communities concerned – a situation that permits even more class-monopoly rents to be realized, since speculators so easily feed off community antagonisms. Parochialist community-based class interest can never be an adequate surrogate for objective social class, for it ignores the essential fact that the survival of the community depends, given the enormous complexity in the division of labor, upon commodity exchanges on a global scale and because it ignores the links in the production and circulation of value in society.

Yet certain kinds of community conflict lead to the formation of nonparochialist horizons. In Baltimore, for example, community groups enraged at the use of the land-installment contract gradually came to realize that financial institutions, by denying conventional mortgage funds while financing speculator-landlords, were the controlling influence in the situation. The community groups began to unravel the skein of argument presented in this chapter through a process of political exploration. And at the end of the road, the community came face to face with what appears to be the dominant power of finance capital.

Curiously enough, there are hints that work-based conflict may lead to the same confrontation. The traditional conflict between worker and industrialist is being ameliorated in certain sectors by the growing integration of workers into management – leading, perhaps, to worker control under certain conditions. But worker control over the factory brings the worker face to face with the power of finance capital to exercise an external control over the activities of industrial enterprise. In much the same way that Marx thought it possible (but unlikely) that land and resources could be brought under state ownership to the advantage of the capitalist, so it appears possible (but unlikely) to nationalize industrial production and to introduce decentralized worker control, without in any way necessarily touching or diminishing the power of finance capital. Worker control has to be viewed, therefore, as a transitional step that fails unless finance capital is also controlled.

The conclusion from the standpoint of both the community and the workplace is that the ultimate power to organize the production and realization of value in society lies in the hands of finance capital. To sustain this conclusion, however, we have to show the necessity for an inner transformation of capitalism such that finance capital comes to exercise a hegemonic power over industrial production as well as all other aspects of life. All that I have space to do here is to provide some clues as to where we might look for the logic of such an inner transformation.

The changing role of money itself provides one such clue. Without money there could be no integrated commodity production, no elaborate division of labor, no price-fixing markets, no universalized exchange values, no medium for the accumulation of capital, no urbanization, and so on. Money in its role of mediator of exchange consequently mediates all significant social interactions. Marx argued that

> the need for exchange and for the transformation of product into pure exchange value progresses in step with the division of labour, i.e. with the increasingly social

character of production. But as the latter grows, so grows the power of money, i.e. the exchange relation establishes itself as a power external to and independent of producers. What originally appeared as a means to promote production becomes a relation alien to producers. As the producers become more dependent on exchange, exchange appears to become independent of them, and the gap between the product as product and the product as exchange value appears to widen. Money does not create these antitheses and contradictions; it is, rather, the development of these contradictions and antitheses which creates the transcendental power of money. (*Grundrisse,* 146)

The "increasingly social character of production" (the increasingly complex division of labor), the constant expansion of capitalist social relations, and the increasing integration of society on a worldwide basis have, since Marx's time, greatly increased "the transcendental power of money." But this power, if it is to be exercised, requires an institutional framework for its expression and a class of people willing and able to make use of it. Marx again provides a clue to the former when he argues that the joint-stock company is an institutional response to the inherent instability of competitive capitalism – an instability that required the concentration, first, of industrial capital, and later, of finance capital. This new arrangement transforms "the actually functioning capitalist into a mere manager, administrator of other people's capital and . . . the owner of capital into a mere owner, a mere money capitalist" (*Capital* 3:436). As a result, interest – "the mere compensation for owning capital that now is entirely divorced from the actual process of reproduction" – is substituted for profit. Marx saw all this creating a transitional mode of production in which new institutions would be increasingly social in character:

This is the abolition of the capitalist mode of production within the capitalist mode of production itself, and hence a self-dissolving contradiction, which *prima facie* represents a mere phase of transition to a new form of production. . . . It establishes a monopoly in certain spheres and thereby requires state interference. It reproduces a new financial aristocracy, a new variety of parasites, in the shape of promoters, speculators, and simply nominal directors. . . . It is private production without the control of private property. (*Capital* 3:438)

Marx did not elaborate much on these remarks, but history has. Industrial corporations have attempted to maintain their independence of financial institutions by generating funds internally, but this has led them to diversify and to take on many characteristics of financial institutions – ITT is almost purely a financial holding company now, and General Motors is steadily moving in that direction. Financial institutions equal and perhaps surpass the industrial corporations in economic power (U.S. House of Representatives 1968, 1971; Herman 1973). State power has grown remarkably and functions to support, by appropriate budgetary, fiscal, and monetary policies, the operations of finance capital. The state is also active in managing both production and consumption (Miliband 1969). Finance capital, operating through state, corporate, and financial institutions, effectively coordinates all social activity into one coherent whole. An industrial capitalism based merely on the immediate production of goods has evolved to a finance form of capitalism which seeks to create and appropriate value through the production, not only of goods, but of new modes of production and new social wants and needs (see Harvey 1982, chaps. 10 and 11). But in doing so, new institutions are founded on the power of money, which is the appearance but not the substance of wealth. Hence arises, in Marx's view, the contradictory character of finance capitalism and its historical necessity as a transitional form.

Financial institutions can accumulate by a variety of techniques. Operating competitively they frequently try to accumulate off each other (by takeovers, asset-stripping, and the like). In aggregate, however, finance capital accumulates out of production in the immediate sense (a work-based exploitation), out of the production of new modes of consumption and the production of new social wants and needs (both of which lead to community-based exploitation). And as finance capital seeks to manage and control the totality of the production process, so there emerges a certain indifference as to whether accumulation takes place by keeping wages down in the immediate production process or by manipulations in the consumption sphere (varying from the manipulation of pension funds to accumulation by means of the processes described in this chapter).

I have already suggested (secs. 3 and 4) that urbanization serves to produce new modes of consumption and new social wants and needs. The roles of speculator-landlord and speculator-developer are crucial to the dynamics of urbanization and therefore to the maintenance of effective demand; and a structure of submarkets through which class-monopoly rents can be realized provides the necessary incentive to play these roles with profit. But at the same time the potential to realize these rents provides the possibility for rapid accumulation of capital out of the land and property markets when the occasion demands it. When industrial demand lags and industrial profits decline, financial institutions will compensate by moving into the land and property markets (ITT has extracted millions out of the Florida real estate boom, for example). But many communities will resist these external forces controlling the conditions of their existence – hence the community conflict typical of advanced urban societies. This analysis suggests a certain underlying unity of community-based and work-based conflict, and herein may lie a clue to the definition of objective classes under advanced urbanization. If objective classes are still to be defined in terms of the production and appropriation of surplus value, then it is now production as a totality (including the production of new modes of consumption and new social wants and needs) rather than immediate production which defines the division between producers and appropriators of surplus value. Marx's theory of surplus value is founded on the analysis of immediate production (with modes of consumption and wants and needs held constant). Exploitation can arise out of the creation of new modes of consumption and the imposition of new social wants and needs – it can impose itself as much on the circulation of revenues as on the circulation of capital (see Chap. 4). The growing hegemonic power of finance capital over the totality of production, circulation, and realization of value in society, therefore, produces a dichotomy between work-based and community-based conflict at the same time as it demonstrates their underlying unity.

This view is reinforced when we turn back to the possibility, broached in section 4, that the processes described in this chapter also serve to generate specific "distributive groupings" or "consumption classes," which in turn define community characteristics in housing submarkets. It is also the case that the production and reproduction of labor power occurs in the community (Giddens 1973, 109–10; Bunge 1971). The reproduction of the social relations of capitalism requires the production of a population which, from the standpoint of employment opportunities and the wages system, will ultimately become fragmented into subjective classes, each prepared to take on certain social roles and to acquire certain technical skills appropriate to its particular position within the overall social structure of a constantly expanding capitalist society. The structure of consumption classes and distributive groupings may, in this fashion, become related to the production of a stratified labor force (see Chap. 5). All urban areas exhibit considerable variation in opportunity to acquire education, social status, social services, and the like (to acquire what Giddens (1973, 103–10) calls "market capacity"). And while there may be considerable individual mobility, it appears that the structure of submarkets which we have identified and the distinctive distributive groupings that occupy them, when combined with the differential distribution of resources to acquire market capacity within the urban system, function to reproduce the social relations of labor under capitalism. These social relations achieve a greater stability precisely because communities, differentiated by social relations, become self-replicating. Objective classes have to be defined, therefore, in terms of a totality of the production process which includes (1) the immediate production of value, (2) the production of new modes of consumption, (3) the production of new social wants and needs, (4) the production and reproduction of labor power, and (5) the production and reproduction of the social relations of capitalism.

VI. FINANCE CAPITAL AND THE URBAN REVOLUTION – A CONCLUSION

We are now in a position to reflect back upon Lefebvre's fundamental thesis. We can provide a comprehensible inner logic for Lefebvre's "ensemble of transformations" through which industrial society comes to be superseded by urban society. In the early years of capitalism, production in particular (the organization of industrial production) was the main focus of attention. In late capitalism, production in all of its facets predictably becomes more and more important. Since the industrialist is adept at immediate production but has little control over the totality of production, finance capital (operating through industrial, financial, and governmental institutions) has emerged as the hegemonic force in advanced capitalist societies. Urbanism has consequently been transformed from an expression of the production needs of the industrialist to an expression of the controlled power of finance capital, backed by the power of the state, over the totality of the production process. Herein lies the significance of urbanization as a mode of consumption and as a producer of new social wants and needs. Concomitantly, the urban realm becomes the locus for the controlled reproduction of the social relations of capitalism. But there also emerges a new definition of objective class interest which is manifest in both work-based and community-based conflicts. In the community these conflicts are over the production of new modes of consumption, new social wants and needs, and the production and reproduction of both labor power and the social relations of capitalism. It seems, however, that the finance form of capitalism, which has emerged as a response to the inherent contradictions in the competitive industrial form, is itself unstable and beset by contradictory tendencies. Of necessity, it treats money as a "thing-in-itself" and thereby constantly tends to undermine the production of value in pursuit of the form rather than the substance of wealth. The alien but transcendental power of money and the institutions created to facilitate the operation of finance capital are not tied to the production of value, and hence we may explain the shift of investment into the secondary circuit of capital at the expense of the primary productive circuit. The perpetual tendency to try to realize value without producing it is, in fact, the central contradiction of the finance form of capitalism. And the tangible manifestations of this central contradiction are writ large in the urban landscapes of the advanced capitalist nations.

The ensemble of transformations of which Lefebvre speaks is far more complex than he imagines. But then so also are the processes of transformation in capitalist society when compared to our ability to grasp them. This complexity cannot be used as an excuse, however, for our almost studied ignorance of the crucial interconnections between the processes of urbanization, economic growth, and capitalist accumulation and the structuring of social classes in advanced capitalist societies. This gap in our thinking is quite odd when the literature on the Third World is so explicit in its dealing with these kinds of relationships. It is rather as if we have succumbed to the illusion that because we are both "advanced" and "urbanized" there is no need to examine the crucial relations through which we arrived at our contemporary state and *which also serve to sustain us in it.*

4

Land Rent under Capitalism

Rent is that theoretical concept through which political economy (of whatever stripe) traditionally confronts the problem of spatial organization and the value to users of naturally occurring or humanly created differentials in fertility. Under the private property arrangements of capitalism, the actual appropriation of land rent by owners forms the basis for various forms of social control over the spatial organization and geographical development of capitalism.

But the social interpretation to be put upon land rent still remains a matter of controversy in the Marxist literature. Marx himself left the topic in a good deal of theoretical confusion. In incomplete and for the most part posthumously published writings, he posed as many conundrums as he solved. The central theoretical difficulty is to explain a payment made to the owners of land (as opposed to improvements embedded by human labor on the land) on the basis of a theory of value in which human labor is key. How can raw land, not itself a product of human labor, have a price (the appearance if not the reality of value)? Marx gives seemingly diametrically opposed answers to this fundamental question. On the one hand he characterizes the value of land as a totally irrational expression that can have no meaning under pure capitalist social relations; on the other he also characterizes ground rent as that "form in which property in land . . . produces value" (*Capital* 3:830–35, 618). In *Theories of Surplus Value* (pt. 2:152) he asserts that if the dominant class relation is between capital and labor then "the circumstances under which the capitalist in turn has to share a part of the . . . surplus value which he has captured with a third, non-working person, are only of secondary importance," whereas in *Capital* (3:618) he discusses how "wage labourers, industrial capitalists, and landowners" together constitute "in their mutual opposition, the framework of modern society" (in itself a startling jump, since landlords suddenly appear as a third "major class" right at the end of an analysis that rests on a two-class interpretation of capitalism).

Any solution to the theoretical conundrum that Marx left behind must be robust enough to handle a wide diversity of practical and material circumstances. Marx himself observed that land can variously function as an *element,* a *means,* or a *condition* of production, or simply be a *reservoir* of other use values (such as mineral resources). Exactly how these different functions acquire political-economic significance depends upon the kind of society we are dealing with and the kinds of activities set in motion. In agriculture, for example, the land becomes a *means* of production in the sense that a production process literally flows through the soil itself. Under capitalism this means that the soil becomes a conduit for the flow of capital through production, therefore a form of fixed capital (or "land capital" as Marx sometimes called it). When factories and houses are placed on the land, then that land functions as a *condition* of production (space), though for the building industry that puts them there in the first place land appears as an *element* of production. Land "demands its tribute" (as Marx puts it) in all of these different senses, but we must also bear in mind that the form and social meaning of rent vary according to these diverse kinds of use. Furthermore, the theory of rent must also encompass a wide diversity of payments – from land-hungry peasants to landlords, from oil-rich potentates seeking prestigious penthouses in the world's capital cities, from industrialists seeking adequate sites for production, from builders seeking land for development, from migrants seeking room and board in the city, from boutique owners seeking access to upper-income clients, and the like. And the landowners are themselves likely to be a motley bunch – wealthy families with large holdings, workers with savings in a small land plot, land companies, churches, insurance companies, banks and mortgage companies, multinational corporations, and the like. The "tribute" that flows on the basis of landownership evidently moves in a multiplicity of directions.

Yet somehow we have to make sense of all this. We desperately need a "scientific analysis of ground-rent," of the "independent and specific form of landed property on the basis of the capitalist mode of production" in its "pure form free of all distorting and obfuscating irrelevancies" (*Capital* 3:624). In what follows I shall attempt such a scientific analysis on the basis of results achieved elsewhere (Harvey 1982, chap. 11). To reduce the levels of confusion, I shall begin with an analysis of land rent appropriated from the circulation of capital in production before proceeding to an analysis of land rent in relation to the circulation of revenues. I shall end with an analysis of land rent in relation to those transitional conditions that typically arise before capitalism is fully implanted as the dominant economic form within a social formation. This permits me to address the problem of so-called "feudal residuals" and the role of land rent in the transition to capitalism. I hope by these steps to arrive at a clearer understanding of the role of land rent in the historical geography of capitalist development.

I. RENT AND THE CIRCULATION OF CAPITAL

The monopoly of private property in land is, Marx asserts, both a "historical premise" and a "continuing basis" for the capitalist mode of production (*Capital* 3:617). Our task is to show how, why, and in what sense this assertion is true. To this end I shall begin with a strong set of simplifying assumptions. First, that all transitional features (feudal residuals) have been eliminated and that we are dealing with a purely capitalist mode of production. Second, that the rent on land can be clearly distinguished from all payments for commodities embodied in the land (land improvements, buildings, etc., which are the product of human labor and which have not yet been fully amortized). Third, that the circulation of *capital* can be clearly distinguished from the circulation of *revenues* (I take up the latter topic in the next section). And, finally, that land has a use value as an element, means, or condition of production (rather than of consumption; note also that I leave aside the concept of land as a reservoir of use values such as mineral resources as a special case). We are then in a position to analyze land rent directly in relation to the circulation of capital.

The best synoptic statement Marx provides of the continuing basis for land rent under capitalism is the following:

> Landed property has nothing to do with the actual process of production. Its role is confined to transferring a portion of the produced surplus value from the pockets of capital to its own. However, the landlord plays a role in the capitalist process of production not merely through the pressure he exerts upon capital, nor merely because large landed property is a prerequisite and condition of capitalist production since it is a prerequisite and condition of the expropriation of the labourer from the means of production, but particularly because he appears as the personification of one of the most essential conditions of production. (*Capital* 3:821)

From this we can distinguish three distinctive roles. The expropriation of the laborer from the land was vital in the stage of primitive accumulation precisely because the land can always be used as a means of production. If labor is to be kept as wage labor then the laborer has to be denied free access to the land. From this standpoint we can see the barrier that landed property puts between labor and the land as socially necessary for the perpetuation of capitalism. This function could just as well be performed, however, if the land becomes state property, "the common property of the bourgeois class, of capital." The problem here is that many members of the bourgeoisie (including capitalists) are landowners, while "an attack upon one form of property . . . might cast considerable doubt on the other form" (*Theories of Surplus Value,* pt. 2:44, 104). From this standpoint, rent can be regarded as a side-payment allowed to landowners to preserve the sanctity and inviolability of private property in general and private ownership of the means of production in particular. This ideological, juridical, and political aspect of landed property is exceedingly important but not in itself sufficient to explain the capitalist forms of rent. The third role of landed property, which turns out to be the most difficult to pin down firmly, is crucial, therefore, to the social interpretation of land rent under capitalism.

The key to the interpretation of the role of landed property under capitalism lies in the pressure it asserts upon the capitalist. The nature of the pressure varies according to the kind of rent extracted. Monopoly and absolute rents interfere with accumulation and arise only to the degree that landed property acts as a barrier to the free flow of capital. Absolute rent, Marx asserts, must eventually disappear (*Capital* 3:765; *Theories of Surplus Value,* pt. 2:244, 393). And monopoly rents, to some degree unavoidable, particularly in urban areas and on land of special qualities (including location), must be kept to a minimum. But absolute and monopoly rents are not the important categories. This conclusion runs counter to Marx's often-quoted assertions concerning the importance of absolute rent (see *Selected Correspondence,* 134) but is consistent with the brief treatments he accords these concepts in *Capital* and *Theories of Surplus Value* compared to the page after page given over to wrestling with the nature of differential rents. I therefore follow Fine (1979) in thinking that Marx's views on differential rent, particularly those partially worked out in *Capital,* are quite distinct from those of Ricardo and provide the clue to the true interpretation to be put upon land rent in relation to the circulation of capital.

Marx follows Ricardo in distinguishing two kinds of differential rent but innovates by analyzing how the two forms of rent relate and "serve simultaneously as limits for one another" (*Capital* 3:737). Marx's insights are hard to recover from chapters full of convoluted argument and elaborate arithmetic calculations. I shall simply summarize the most important features.

Differential rent of the first type (DR-1) arises because producers on superior soils or in superior locations receive excess profits relative to production costs on the worst land in the worst locations. Superior soils and locations, like superior technology, are indeed sources of relative surplus value to individual producers (which explains why all of them can appear as "productive of value"). But unlike superior technology, superior locations and soils can form relatively permanent sources of excess profits. If the latter are taxed away as rent, the profit rate is equalized across different soils and locations. Capitalists can then compete with each other only through adoption of superior techniques – which pushes the capitalist system back onto its central track of looking to revolutions in the productive forces as the means to its salvation. The extraction of DR-1 has a vital social function in relation to the dynamics of capital accumulation. Without it, some producers could sit complacently on the excess profits conferred by "natural" or "locational" advantages and fail in their mission to revolutionize the productive forces on the land.

This conception of DR-1, essentially no different from that of Ricardo, has to be modified in three important respects. First, trade-offs can exist between fertility and location so that the worst land has to be understood as a combination of characteristics. Second, both fertility and location are social appraisals and subject to modification either directly (through soil exhaustion or improvement, changing transport facilities, etc.) or indirectly (through changing techniques of production which have different land or locational requirements). The excess profits from superior soils or locations are permanent only in relation to changing appraisals. Third, DR-1 depends upon a "normal" flow of capital into production on the land. And when we switch to consider what constitutes that normal flow of capital onto the land, we immediately encounter the problem of the second kind of differential rent (DR-2). The immediate implication is that DR-1 depends crucially on capital flows that automatically generate DR-2.

Imagine a situation in which no advantages due to location or fertility existed. Differentials in productivity on the land would then be caused solely by the different quantities of capital invested (assuming some pattern of returns to scale). Excess profits in this case are entirely due to the investment of capital. Conversion of these excess profits into DR-2 will simply check the flow of capital onto the land except under two particular conditions. First, if the investments embed relatively permanent productive forces in the land, then the flow of capital leaves behind a residue of improvements which form the basis for the appropriation of DR-1. Such residues (drained and cleared land and other forms of land improvement) are widespread and of great importance. Second, the direct appropriation of DR-2 can, under the right circumstances, prevent the flow of capital down channels that might be productive of profit for the individual capitalist but that would have a negative impact upon the aggregate growth in surplus value production. We here encounter a classic situation in which individuals, left to their own devices and coerced by competition, would engage in investment strategies that would undermine the conditions for the reproduction of the capitalist class as a whole. Under such conditions the external discipline imposed by landowners, like the external discipline exercised through the credit system, has a potentially positive effect in relation to the stabilization of accumulation. The emphasis, however, has to be upon the "potentiality" of this result, because what Marx's tedious arithmetic examples appear to show is that the appropriation of DR-2 can exercise a negative, neutral, or positive pressure upon the accumulation of capital depending upon the circumstances. Furthermore, the flow of capital onto the land also depends upon general conditions of accumulation – a plethora of capital in general or particular conditions prevailing within the credit system have direct implications for the flow of capital onto the land (*Capital* 3:770, 676).

Now combine the interpretations of DR-1 and DR-2. We will be seriously in error, as Fine (1979) points out, if we treat the two forms of rent as separate and additive. Insofar as DR-1 depends on a social appraisal of "natural and locational advantage," it depends upon capital flows that often modify nature in crucial ways. The appropriation of DR-2, for its part, could not occur without DR-1 as its basis. The two forms of differential rent in effect merge to the point where the distinction between what is due to land (with the aim of equalizing the rate of profit and keeping the impulsion to revolutionize the productive forces engaged) and what is due to capital (with the aim of keeping the flow of capital into revolutionizing the productive forces on the land at a level consistent with sustained accumulation) is rendered opaque. In other words, the appropriation of rent internalizes contradictory functions. The permanent tension between landowners and capitalists within a purely capitalistic mode of production is a reflection of this contradiction. Furthermore, to the degree that rental appropriation can have negative, neutral, or positive effects in relation to accumulation, the social relations that arise in response to this contradiction can have a powerful effect upon the allocation of capital to land, the whole structure of spatial organization, and, hence, the overall dynamic of accumulation. In order to explore these contradictions and their effects, however, we must first establish the form that private property in land must assume if it is to be integrated within a purely capitalist mode of production.

The conclusion to which Marx points, without full explication, is that land must be treated as a pure financial asset and that land has to become a form of "fictitious capital." This conclusion calls for some explanation (Harvey 1982, chaps. 9 and 11). "Fictitious capital" amounts to a property right over some future revenue. Stocks and shares, for example, can be sold before any actual production takes place. The buyers trade their money in return for a share of the fruits of future labor. Insofar as the money is used to set labor in motion (or create conditions, such as physical infrastructures, to enhance the productivity of social labor), then the fictitious capital stands to be realized. Even under the best conditions, fictitious capital entails speculation; and under the worst, it provides abundant opportunity for fraud and devaluation. Capitalism could not function, however, without the large-scale creation and movement of fictitious forms of capital via the credit system and capital markets. Only in this way can capital be shifted rapidly from unprofitable to profitable sectors and regions, new lines of activity be opened up, centralization of capitals be achieved, etc. The credit system (capital markets in particular) becomes the central nervous system for the coordination of accumulation. It also becomes the central locus of all of capitalism's internal contradictions. Crises always appear, in the first instance, therefore, as financial and monetary crises.

Once the significance of fictitious forms of capital is established, we can see how property rights over any form of future revenue might be bought and traded. Government debt (a right to a share of future taxes) and mortgages on land (a right to future rents) and property (a right to amortization payments) all stand to be freely traded. In the case of land, what is bought and sold is the title to the ground rent yielded by it. That ground rent, when capitalized at the going rate of interest, yields the land price. Hence arises an intimate relationship between rent and interest. The money laid out by the buyer of land is equivalent to an interest-bearing investment, a claim upon the future fruits of labor. Title to land becomes a form of fictitious capital, in principle no different from stocks and shares, government bonds, etc. (although it has certain qualities of security, illiquidity, etc.). Land, in short, can be regarded as a pure financial asset. This is the condition, I argue, that dictates the pure form of landed property under capitalism.

The theory of ground rent tells us that landowners should ruthlessly appropriate all excess profits due to relatively permanent advantages of fertility or location (no matter whether the product of capital or not). Anticipated future excess profits (due to future capital flows and future labor) affect the price of land in the present insofar as land becomes a pure financial asset, a form of fictitious capital. Marx excluded such speculative activity from his purview (except in a few rare instances, e.g., *Capital* 3:774–76) and was therefore content to view landownership as an entirely passive function. But land markets, like capital markets, play a vital coordinating role in the allocation of future capital and labor to the land. Landowners leave behind their passive stances and can play an active role in the creation of conditions that permit enhanced future rents to be appropriated. In so doing, of course, they condemn future labor to ever-increasing levels of exploitation in the name of the land itself. But they also play a vital role in relation to accumulation.

Landowners can coerce or cooperate with capital to ensure the creation of enhanced ground rents in the future. By perpetually striving to put the land under its "highest and best use," they create a sorting device that sifts land uses and forces allocations of capital and labor that might not otherwise occur. They also inject a fluidity and dynamism into the use of land which would otherwise be hard to generate and so adjust the use of land to social requirements. They thereby shape the geographical structure of production, exchange, and consumption, the technical and social division of labor in space, and the socioeconomic spaces of reproduction, and invariably exert a powerful influence over investment in physical infrastructures (particularly transportation). They typically compete for that particular pattern of development, that particular bundle of investments and activities, which has the best prospect for enhancing future rents. In this way, as Marx puts it, "rent, instead of binding man to Nature, merely bound the exploitation of the land to competition" (*Poverty of Philosophy,* 159).

We can now bring the argument full circle. Not only is the appropriation of rent socially necessary under capitalism by virtue of the key coordinating functions it performs, but landowners must also treat the land as a pure financial asset, a form of fictitious capital, and seek, thereby, an active role in coordinating the flow of capital onto and through the land. The effect is to free up the land to the circulation of interest-bearing capital and to tie land markets, land uses, and spatial organization into the general circulation process of capital.

But by the same token, the more open the land market is, the more recklessly can surplus money capital build pyramids of debt claims and seek to realize these claims through the pillaging and destruction of the land itself. Investment in appropriation, so necessary if the land market is to perform its vital coordinating functions, simultaneously opens up the land to "all manner of insane forms" let loose within the credit system in general. What appears as a sane and sober device for coordinating the use of land to surplus value production and realization can all too easily dissolve into a nightmare of incoherence and periodic orgies of speculation. Here, as elsewhere, the only ultimate form of rationality to which capitalism responds is the irrationality of crisis.

I can, at this point, rest my case. There is a form of landownership and land rent which fully integrates with the circulation of capital. Land markets, like capital markets, do not produce value in the primary sense, but they play a vital role in coordinating the application of social labor. Capitalism simply would not work without them. And land markets could not exist without land rent, the appropriation of excess profit from capital. The crisis-prone character of capitalism is, of course, carried over and even heightened within the credit system as well as within land markets. The detailed study of the specific form these internal contradictions take within land markets is an urgent matter.

Yet all of this requires that land be treated as a pure financial asset, a form of fictitious capital. This requires that the power of any distinct class of landowners be broken, that ownership of land become from all standpoints (including psychological) simply a matter of choosing what kinds of assets to include in a general portfolio of investments. And this, of course, is increasingly how pension funds, insurance companies, and even private individuals tend to view land investment. This is not to say that in practice all traditional forms of landownership have disappeared in the advanced capitalist world. But it is interesting to note that land has long been dominantly viewed as a pure financial asset in the United States (the country least hindered by "feudal residuals") and that the direction of the transition in countries such as Britain has been very much toward the acceptance of land as a form of fictitious capital. The point, of course, is that these forms of landownership (and the social attitudes they generate) are entirely consistent with the circulation of capital, at the same time as they are fully expressive of the internal contradictions within that circulation process.

II. RENT AND THE CIRCULATION OF REVENUES

Within a purely capitalist mode of production, all forms of revenue – wages, profit of enterprise, interest, taxes, rent, etc. – have their origin in the production of value and surplus value. But, once distributed, revenues are free to circulate, thereby creating opportunities for various secondary forms of exploitation. Rents can, therefore, just as easily be appropriated from the circulation of revenues as from the circulation of capital. Landlords are presumably indifferent to the immediate origin of the rental payment. They are satisfied as long as the rent keeps rolling in. But the theoretical distinctions are of interest because circumstances often arise, particularly though not exclusively in urban areas, where it is impossible to understand the social meaning of rental payments without explicit consideration of the circulation of revenues.

The intricacy of the circulation of revenues bears some initial elaboration. The total wage bill, for example, is split among different factions of the proletariat according to their reproduction costs and their gains through class struggle. Capitalists, furthermore, do not normally discriminate as to individual needs of their workers and simply pay the going rate for a job. But individual worker wants and needs vary enormously depending upon family status, age, health, and, of course, tastes and fancies. On a given day, therefore, some workers will possess money surpluses while others will be unable to meet their needs. The stage is set for the circulation of wage revenues within the working class. First, payments may be made for services rendered (baby-sitting, washing, mending, cooking, etc.). Second, workers may borrow and lend to one another, sometimes at a rate of interest. The early benefit societies, savings and loan associations, and so forth, were simply attempts to institutionalize such activity. The extent of this circulation of revenue varies, but it can be quite massive. The social security system, for example, is a transfer from those now working to those now retired, in return for a claim on a share of future wages.

Various factions of the bourgeoisie can also circulate revenues amongst and between one another, either by trading services or through intricate patterns of lending (read any tale by Balzac or Dickens to get some sense of the importance, socially, of this form of the circulation of revenues). Revenues can also circulate between the social classes. The bourgeoisie frequently hires workers at a going rate to render the kinds of services that taste, fashion, custom, fancy, and income dictate. The cook, the valet, the prostitute, the gardener, are all paid out of the circulation of revenues and not directly out of the circulation of capital. Workers may likewise pay, or be forced to pay, for services rendered them by the bourgeoisie (legal, administrative, etc.).

The circulation of revenues is, evidently, both intricate in its detail and massive in scale. Much of the detail of what happens in bourgeois society, including the appropriation of rent, has to be understood in relation to it. We will be seriously in error if we seek to explain that detail by direct reference to categories Marx designed to cope with the dynamics of capital circulation. Yet the circulation of revenues necessarily integrates with the circulation of capital at well-defined points. In a purely capitalist mode of production, all revenues have their origin in value and surplus value production and ultimately return to the circulation of capital through the purchase of commodities. If this aggregate relation does not hold, then the circulation of capital breaks down. Furthermore, to the degree that circulation time is vital to capital, time lost through the circulation of revenues is a drag upon accumulation. But we here encounter situations in which an appropriate circulation of revenues can play a positive role in relation to accumulation. If, for example, houses had to be bought for cash and individuals had to save the full cash amount to buy them, then vast sums of money would have to be hoarded. The circulation of money as capital and the demand for housing (an important field for accumulation) would both be held in check. Renting or a credit system through which individuals with surplus savings can lend to those in need overcomes the blockage. Hoards are reduced and the free circulation of capital maximized. An inadequate structure for the circulation of revenues can therefore act as a barrier to the circulation of capital. State policy, particularly in its welfare aspects, has often been dedicated to achieving more efficient structures for the circulation of revenues in relation to the circulation of capital. There is, in this, a certain convenience because social unrest can often be coopted by reforms that appear to satisfy worker needs by rationalizing the circulation of revenues while leaving the circulation of capital if anything enhanced rather than diminished. The reforms of the American New Deal were very much of this sort.

The problem, of course, is that once revenues begin to circulate, then there is nothing to guarantee that they will do so in ways appropriate to the circulation of capital. While lending to cover the purchase of a house may appear perfectly rational from the standpoint of accumulation, the passing of

IOU's to cover spiraling gambling debts does not. All kinds of opportunities exist, also, for secondary forms of exploitation – usurious lending or rack-renting practices, for example. And to cap it all, the distinction between the circulation of capital and the circulation of revenues is also hard to sustain in the diverse money transactions that characterize daily life. If workers lend to other workers at a rate of interest, then why can't they lend also to capitalists, particularly if the rate of interest is higher? And if workers borrow, then what is to stop the rise of usurious practices within the working class (the hated pawnshop) or the penetration of capitalist lending to control and stimulate working-class consumption? The problem arises here because the distinction between workers and capitalists is obscured within the credit system where the primary relation is between lenders and borrowers, debtors and creditors, of whatever sort. All kinds of cross-currents then arise inconsistent with the primary forms of exploitation that Marx dwells upon at such length and so exclusively.

Consider, now, how this relates to the social interpretation to be put upon the rental payment. The monopoly power conferred by private property in land always remains the basis. But we now see that rent is not only appropriated from capitalists as a straight deduction out of the surplus value produced under their command. Rent is also levied from workers, other members of the bourgeoisie (financiers, professionals, retired businessmen, other landlords, etc.), the state, and cultural, religious, and educational organizations. Rent can be appropriated from the circulation of revenues of *all* sorts. We can hardly use Marx's categories (which deal solely with the circulation of capital) to explain the rent paid by company executives for penthouses in Paris or London, the rent paid by rich retirees in Florida or unemployed blacks in Baltimore's ghetto. Furthermore, to the degree that land has been reduced to a pure form of fictitious capital, a pure financial asset, anyone who saves can invest in it, appropriate rent, and speculate in land price. If workers own small plots of land and the property thereon, they can just as well play this game as anyone else. Indeed, petty land and property trading and renting (even a room) has been a prime means for upward mobility within the working class and the petite bourgeoisie for centuries. The destruction of the singular power of any coherent landlord class, a concomitant of the rise of the purely capitalist form of ground rent (see Massey and Catalano 1978), opens up the possibility to anyone who has savings to invest them in land and so acquire the power to appropriate rent.

Is there any way to see through this crazy patchwork quilt of social relations and say something coherent about the social meaning of the rental payment? I think Marx's analysis, suitably modified, remains helpful.

To begin with, we can situate the appropriation of rent as a moment within the circulation of revenues in general and then invoke the necessary relation between the latter and the circulation of capital in order to explore the limits of the total capacity for rental appropriation. If all surplus value is appropriated and held as rent, then there would be no place for the accumulation of capital or, for that matter, any other form of revenue circulation. I quote the *reductio* case to pose this question: how much rental appropriation is appropriate to sustain the accumulation of capital? We have already argued that rental appropriation from production has an important function to play in allocating land to uses and in shaping spatial organization to capitalism's requirements. We now see that exactly the same consideration holds for the "rational" allocation of consumption uses across space. The flow of fictitious capital through the land therefore has the positive attribute that it can forge "rational" spatial configurations of both production and consumption in relationship to aggregate accumulation. Capitalism stands to benefit from the persistence of private property and land rent. The problem, of course, is that there is nothing to prevent the rise of all manner of insane speculative and monopolistic practices within the field of rental appropriation or the transmission of speculative impulses from within the credit system. The dual consequence of excessive appropriation of rent in relation to the circulation of capital and gross distortions of spatial structure leads to strong demands to eliminate or control the power to appropriate rent through state interference. The positive, neutral, or negative effects of rental appropriation in relation to accumulation remain perpetually with us, part and parcel of the equilibrating and disequilibrating tendencies within the capitalist mode of production.

The interior details of the appropriation of rent in relation to the circulation of revenues are also open to closer scrutiny, in part by analogy to the basic Marxian categories. Consider, for example, the rent appropriated from the working class. A secondary form of exploitation, as both Marx (in passing) and Engels (1935) argued, the rent extracted from the worker can affect the value of labor power and so diminish surplus value to the capitalist. On this basis Engels attacked those who sought any solution to the housing question in the absence of any attack upon the wages question. While correct in substance, this does not absolve us from the need to analyze the economic, social, and ideological consequences of the rental payment. Analysis of such questions reveals some interesting insights. For example, if workers receive a uniform wage, then those who live close to work will incur lower transport costs and therefore lower costs of social reproduction. Properly structured rents on working-class housing would then have the effect of equalizing the real wage to workers at different locations. The analogy with differential rent on capital is exact. The problem, of course, is that there is nothing in the power relations between landowners and workers to ensure that rents are "properly" structured in the first place. Furthermore, workers also compete for living space against capitalist producers and bourgeois consumers. The level of appropriation of rent from one kind of revenue cannot be understood independently of the others. The relations between land and property rent, transport availability, employment opportunities, and housing, as well as other consumer functions, all in the context of shifting geographical patterns in the circulation of both capital and revenues, then define the nexus of forces shaping the spatial configuration of land uses.

It is into such a situation that those who use land as an *element* of production – the developers and builders – insert themselves as the prime movers creating new spatial configurations of the built environment and new opportunities for rental appropriation. The analogy is with DR-2, but in this case the investment of capital yields its return through an enhanced capacity to tap the circulation of revenues. This is as true for the builders of back-to-back housing as it is for the developers of expensive condominiums for the haute bourgeoisie. Fictitious capital in land makes its claim upon future labor indirectly (as in the case of housing purchase with the aid of a mortgage) through the future circulation of wages and other forms of revenue.

But the combination of the monopoly privileges inherent in any form of private property in land and the active processes of production of particular spatial configurations generates many an opportunity to garner monopoly rents. This tendency, as Marx observed, is particularly strong in urban areas. Specific sites can command a premium land rent precisely because of their privileged location relative to previous investments. Indeed, whole islands of privilege can be constructed within which all landowners acquire the collective power to garner monopoly rent – be it landlords within the confines of the ghetto or developers peddling loft space to affluent young professionals in New York's Soho district (Zukin 1982). Situations arise, therefore, in which the concept of "class-monopoly rent" (see Chap. 3) makes eminent sense. We do not have to appeal here to the idea of any inherent class power on the part of all landholders nor even depart from the concept of land as a pure financial asset, a form of fictitious capital. We simply have to recognize that within the complex matrix of urban development, situations arise in which space can be collectively monopolized and a given pattern of the circulation of revenues trapped within its confines. Even the concept of "housing class" makes sense when projected and understood against such a background.

The full theory of such relationships remains to be worked out. Marx's theory of rent is partial because it deals solely with the circulation of capital and excludes any direct analysis of the circulation of revenues. We cannot, therefore, simply take Marx's categories and make them work for us in the actual analysis of the total complex of land and property markets (particularly in urban areas). Something more is involved, even presuming a purely capitalist mode of production. And that something more is the circulation of revenues, albeit always related back to the circulation of capital that necessarily lies at its basis. A closer look at how rent is appropriated from the circulation of both capital and revenues will generate more precise understandings of the surface appearance of the functioning of land markets without abandoning the deep structural insights that Marx generated.

III. RENT IN TRANSITIONAL SOCIAL FORMATIONS

The social interpretation to be put on rent varies from society to society and undergoes a fundamental alteration with the transition from one mode of production to another. "In each historical epoch," Marx writes, "property has developed differently and under a set of entirely different social relations" (*Poverty of Philosophy,* 154). We will be grossly in error if we interpret feudal rents, or rents in the transitional phase when landed capital held sway, directly by reference to the role of rent in advanced capitalist society. Yet a knowledge of the latter is indispensable to an interpretation of the former. Furthermore, it is in the nature of the transition to merge two often quite antagonistic roles so they become indistinguishable. The difficulty is, then, to keep the social interpretations distinct from each other while simultaneously understanding how they can coexist within the same money payment for the use of land. Only in this way can we understand how one form of rent is gradually converted into the other through a material historical process.

Marx considered that money rent on land and its corollary, the formation of land markets, were preconditions for the rise of capitalism. Like merchants' capital and usury, landed capital precedes the modern standard form of capital. The latter ultimately subjugates these earlier forms and converts them to its own requirements. The actual history of this process is strewn with complexities generated out of the cross-currents of class struggle and the diversity of initial conditions of land tenure and ownership. Marx's general version of this history, based on the Western European experience, can be divided into two phases. In the first, feudal labor rents (the source of a surplus product) are transformed into rent in kind and finally into a money payment, while land is increasingly released from those constraints that prevent it from being freely traded as a commodity. Furthermore, the conversion to money payments implies either a voluntary or a forcible integration of land users (particularly agricultural producers) into some kind of general system of commodity production and exchange.

None of this ensures, however, that rent assumes its modern, purely capitalistic form, thoroughly integrated into the circulation of capital (and revenues). All kinds of intermediate forms can arise. In a stimulating and provocative work, Pierre-Philippe Rey (1973) proposes that these be viewed as "complex articulations" of different modes of production, one upon the other. Rey goes on to show how material conditions and class configurations and alliances can freeze the transition in a half-way state between precapitalist and capitalist modes of production for extended periods (as often appears to be the case in Third World peasant societies). What Marx viewed as an inevitable though lengthy transition can, in Rey's view, be all too easily blocked. Unfortunately, Rey goes on to argue that rent has no real basis within a capitalist mode of production and that it can be interpreted only as a relation of distribution which reflects a relation of production of precapitalist modes of production (e.g., feudalism) with which capitalism is "articulated." We have already seen how that conclusion can be refuted and a real social basis uncovered for the appropriation of rent within a purely capitalist mode of production. But for the moment we shall follow Rey's perceptive line of argument as it applies to the transitional phase.

Under transitional conditions, landlords can play a direct and active role in the exploitation of labor (as opposed to the backseat, passive role that Marx [incorrectly] assigned them under capitalism). This is as true for slave economies (the American South prior to the Civil War) as it is for landlord-peasant systems of agricultural production in the present era. There is a direct incentive for the landlord to extract the maximum of rent (whether in kind or in money does not immediately concern us), not only because this maximizes the landlord's revenues, but also because the peasant is forced to work harder and harder and produce more and more commodities for the market at ever lower prices (given the increase in supply). The massive exploitation of a rural peasantry by a landlord class is, from this standpoint, perfectly consistent with industrial capitalism as long as it provides cheap food for workers and cheap raw materials for industry. And even if the peasants nominally own their own land (there is no overt landlord class), indebtedness at usurious interest rates and the obligation to pay taxes to the state can have the same effect. It is not hard to see how a powerful alliance of classes comprising landowners, industrial bourgeoisie, and money lenders, backed by the state, can form and block any full transition to capitalist social relations on the land.

But such a form of exploitation, like absolute surplus value (of which it is a bastard form), has negative social consequences and inherent limits. First, the extraction of a fixed money payment from basically subsistence producers may diminish the supply of commodities when prices rise because producers have to sell less to reach a fixed money goal. Prices continue to rise as a consequence. When prices go down, however, peasants have to sell more and so increase the supply in the face of falling prices. Price movements and commodity supplies do not, under such conditions, integrate at all well with the general dynamic of accumulation. That dynamic, second, invariably requires an expansion of output which, with a fixed technology of peasant production, means increasing rates of exploitation. And as exploitation increases, so do the conditions for revolutionary movements ripen. Even in the absence of such resistance, there is an absolute limit to this kind of absolute exploitation. At some point the productive forces on the land have to be revolutionized to accommodate the expanding demands of capitalism. We then discover that the transitional forms of organization inhibit "the development of the social productive forces of labour, social forms of labour, social concentration of capital . . . and the progressive application of science" (*Capital* 3:807). New productive forces have to be deployed, and that means opening up the land to the free flow of capital.

This brings us to the second phase of Marx's version of the transition to capitalist forms of rental appropriation. Capital and labor must confront each other on the land free from any direct interference of the landlord class. The landlord must be reduced to a purely passive figure. The class alliance between an industrial bourgeoisie and a landlord class breaks down, and antagonistic relations arise between them until such time as the latter class entirely disappears as a coherent force in society. And all of this must happen because this is the only way, under capitalism, for productive forces to be revolutionized on the land itself.

We can understand this transition from the standpoint of the landlord in the following way. The landlord can dominate a peasantry tied to the land and has everything to gain from maximizing the extraction of rent. But the landlord cannot similarly compel the capitalist to invest, and therefore has much to lose from maximizing the extraction of rent from the capitalist. The power of landed property then acts as a barrier to the free flow of capital onto the land and inhibits the development of the productive forces. The possibility exists, however, for a terrain of compromise between landowner and capitalist. The use value of land to the capitalist is as an element, means, or condition of production which, when worked on by labor, produces surplus value. The capitalist is concerned with rent in relation to surplus value produced. The landlord, in contrast, is concerned with the rent per acre. Under conditions of strong capital flow onto the land the rent per acre can rise while the rent as a portion of surplus value produced declines (cf. *Capital* 3:683). Under these conditions, the landlord has everything to gain by minimizing the barrier that landed property places to the flow of capital. This was, of course, the basis of the compromise that existed in England during the period of "high farming," 1850–73.

The relationship between capital and landed property is not reduced thereby to one of perpetual harmony. It is often hard to distinguish, for example, between peasant producers and small-scale capitalist producers, while landlords may not be sophisticated enough to appreciate the long-term gain to them of the shift from rack-renting peasants to seducing capitalists to invest. Also, to the degree that the expansion of social labor "stimulates the demand for land itself," landed property acquires "the capacity of capturing an ever-increasing portion" of surplus value (*Capital* 3:637–39). Blessed with such an opportunity, what landlord could resist exploiting it? The landlord is perpetually caught between the evident foolishness of extracting too little from capital and the penalties that accrue from trying to take too much. And there are, in addition, all kinds of institutional problems relating to permanent improvements, tenancy conditions, leasing arrangements, and the like, which are the focus of interminable struggles between capitalist and landlord. Like contractual issues that arise between capital and labor these institutional arrangements are ultimate regulated through the state.

Marx evidently did not feel too secure in his rendition of how the capitalist form of private property came to be. He was later to claim that he had merely sought to "trace the path by which, in Western Europe, the capitalist economic system emerged from the womb of the feudal economic system," and he attacked those who transformed his "historical sketch of the genesis of capitalism into an historico-philosophical theory of the general path of development prescribed by fate to all nations, whatever the historical circumstances in which they find themselves" (*Selected Correspondence,* 312–13). His studies of the evolution of landed property in the colonies and the United States, as well as in Russia, convinced him that the transition was not unilinear. Even in Western Europe considerable variation existed, in part because of residual features "dragged over into modern times from the natural economy of the Middle Ages," but also because of the uneven penetration of capitalist relations under historical circumstances showing "infinite variations and gradations in appearance" which demand careful empirical study (*Capital* 3:787–93). Under such conditions even the neatness of the two-phase transition breaks down. We almost certainly will find radically different forms of rental appropriation side by side.

Properly used, Marx's framework can provide many insights. For example, in a careful reconstruction of the historical record of rental appropriation in the Soissonais district of France, Postel-Vinay (1974) shows that over the last two centuries large-scale farmers working the better land have consistently paid about half the rent per acre extracted from small peasant proprietors working inferior soils. In Rey's eyes this makes a mockery of both Marxian and neoclassical views of differential rent (supposedly extracted from lands with superior productivity) and confirms his view that rent can be understood only as a feudal relation of production perpetuated as a capitalist relation of distribution. Rey is only partly right. If my reading of Marx is correct, then the superior rent paid by small peasant proprietors is a reflection of a landlord-labor relation as opposed to the landlord-capital relation found on the better lands. Two different social relationships have coexisted within the same region for two centuries. Yet rent is still paid by the capitalists according to a logic that has nothing to do with the articulation of feudal and capitalist modes of production. Rey's depiction of conditions during the transition (including blocked transitions that freeze social relations into the landlord-laborer pattern) may be quite appropriate. But he is way off target when he asserts that this is the only form that rent can take under capitalism.

The possibility that radically different social relations may coexist within a given region over extended periods must give us pause. It alerts us to the danger of assuming that the same social interpretation can be put upon the rental payment even within seemingly coherent capitalist social formations. This is not to make any strong claim for the persistence of "feudal residuals" under capitalism. It simply means that the owners of any important means of production (land, productive capacity, money) have the habit of trying to appropriate as much surplus value as they can by virtue of that ownership and that circumstances have to be very special to reduce them to that "passive" state that Marx depicted. Furthermore, as I initially argued, there is an active role for rental appropriation even under the purest form of capitalism.

IV. CONCLUSION

Rent is, I repeat, simply a money payment for the use of land and its appurtenances. This simple money payment can conceal a host of possible social significations that can be unraveled only through careful sociohistorical investigation. The task of theory under such circumstances is to establish the underlying forces that give social meaning to and fix the level of the rental payment. Under a purely capitalist mode of production, these forces merit disaggregation into those that attach to the circulation of capital and those that relate to the circulation of revenues (while recognizing that the two circulation processes are dependent upon each other). Additional complications arise because it is not always easy to distinguish between interest on capital fixed in land (interest on buildings and permanent improvements) and rent on land pure and simple. Furthermore, the different uses of land as a means, condition, or element of production, or as a reservoir of present or potential use values, means that the significance of land to users varies from sector to sector. Project all of these complexities into the framework of land use competition in which land is just one of several different forms of fictitious capital (stocks and bonds, government debt, etc.) competing for investment, and we are forced to conclude that there is nothing simple about that simple money payment even under conditions of a purely capitalist mode of production.

But the notion of a purely capitalist mode of production is at best a convenient fiction, more or less useful depending upon historical circumstance. And many situations indeed arise in which the dominant forces underlying the rental payment can best be understood in terms of the articulation of quite different modes of production, one upon the other. To the degree that different modes of production have specific forms of distribution and revenue circulation associated with them, all of the complexity of a purely capitalist mode of production becomes compounded many times over.

I do not regard the rich complexity of these theoretical determinations as anything other than an exciting challenge to bring the theory of rent out from the depths of underlying simplicity (where some Marxists seek to confine it in perpetuity) step by step toward the surface appearance of everyday life. The framework outlined here can have as much to say about the role of private property in land and the appropriation of rent in the social transformation of Kinshasa as it can be used to look at landlordism in Baltimore, loft-living and gentrification in New York, landed property in the Soissonais, and corporate farming in Iowa. The theory does not tell us the answers, but it does help us pose the right questions. It might also help us get back to some basic issues about class structures and alliances, different modes of appropriation and exploitation, and the role of landownership as a form of social power in the shaping of spatial configurations of land uses. And all of this can be done, I would submit, without refuting or "going beyond" supposedly outdated Marxian formulations but simply through the proper application of Marx's own methods to a question that he never himself resolved to his own satisfaction.

5

Class Structure and the Theory of Residential Differentiation

The theory of residential differentiation is desperately in need of revision. Sociological explanations of residential differentiation (see the review by Timms 1971) have never progressed much beyond elaborations on the rather simplistic theme that similar people like to, or just do, live close to each other. The seeming complexity of sociological accounts derives from the difficulty of defining "similar" and the difficulty of showing whether people are similar because they live close to each other or live close to each other because they are similar. The explanations constructed out of neoclassical economic theory are no less simplistic in that they rely upon consumer sovereignty and utility-maximizing behavior on the part of individuals which, when expressed in the market context, produces residential differentiation. Complexity in this case arises because it is not easy to give concrete meaning to the utility concept and because it is possible to envisage a wide variety of conditions under which individuals might express their market choices.

Most thoughtful commentators on the matter have concluded that the problem lies in specifying the *necessary* relationships between social structure in general and residential differentiation in particular. For example, Hawley and Duncan (1957, 342) remark that "one searches in vain for a statement explaining why residential areas would differ from one another or be internally homogeneous. The elaborate discussion of social trends accompanying urbanization is nowhere shown to be relevant to this problem." Most attempts to integrate social theory and the theory of residential differentiation have produced, in fact, "not a single integrated theory, connecting residential differentiation with societal development, but, rather, two quite distinct theories which are accidentally articulated to the extent that they happen to share the same operational methods."

The problem here lies in part in the realm of methodology. Plainly, it is inappropriate to speak of residential differentiation causing or being caused by changes in the total social structure, while a functionalist language, although somewhat more appropriate, is so dominated by the notion of harmonious equilibrium that it cannot deal with the complex dynamics and evolutionary character of a capitalist society. Yet most analysts have been trapped into the use of an inappropriate causal or functionalist language when they have dared to venture beyond statistical descriptions. All of this has produced an enormous amount of material on a variety of facets of the residential differentiation process, but no clue is provided as to how this material might be integrated into general social theory.

The Marxian method, however, founded in the philosophy of internal relations (Ollman 1971), is fashioned precisely to provide a coherent methodology for relating parts to wholes and wholes to parts. Indeed, the central conception in Marx's version of the dialectic was to view things relationally in order that the integrity of the relationship between the whole and the part should always be maintained. Consequently, Marx criticized the categories of bourgeois social science on the grounds that they are abstractly fashioned without reference to the "relations which link these abstractions to the totality" (Ollman 1973, 495). Marx's abstractions are of a different kind, for they focus on such things as social relations. Relatively simple structures might be isolated from the whole for purposes of analysis, but "what is decisive is whether this process of isolation is a means towards understanding the whole and whether it is integrated within the context it presupposes and requires, or whether the abstract knowledge of an isolated fragment retains its 'autonomy' and becomes an end in itself" (Lukács 1968, 8).

The theory of residential differentiation has rarely been subjected to an analysis from a Marxian standpoint, and it is predictable, therefore, that the "theory" consists of an incoherent mass of autonomous bits and pieces of information, arrived at by means of studies each conceived of as an end in itself and each conceived in terms of relationships specified in a causal, functional, or empiricist language (with all the limitations that each of these imposes). And it is predictable that attempts to integrate this material into some general social theory would meet with little or no success. In this chapter I shall therefore attempt an outline of the relation between residential differentiation and social structure. Such an investigation is bound to be preliminary and sketchy at this stage. But I hope to show where the key relations lie and thereby indicate where we have to look for a revision of the theory of residential differentiation that will make sense. I shall begin with an analysis of the forces creating class structure in advanced capitalist society.

I. CLASS AND CLASS STRUCTURE

Theories of class and class structure abound. Marx and Weber laid the basis, and a host of contemporary interpreters have added insights, glosses, reinterpretations, and, it must be added, mystifications. Rather than attempt a synthesis of this work, I shall sketch in a theory of class which derives primarily from a reading of Marx and secondarily from adapting materials from Giddens (1973) and Poulantzas (1973).

A central tenet of Marx's historical and materialist method is that a concept such as "class" can take on a meaning only in relation to the historical context in which it is to be applied. "Class" has a contingent meaning depending upon whether we are considering feudal, capitalist, or socialist modes of production. Class theory is not therefore, a matter of identifying a fixed set of categories which are supposed to apply for all times and places. The relational view of class which Marx espouses focuses our attention on the forces of "class structuration" (as Giddens 1973 calls them) which shape actual class configurations. In the context of the capitalist mode of production, however, "class" has a more specific meaning that relates to the basic social relationships pertaining in capitalist society. The forces of class structuration under capitalism are identical to those contained in the dynamics of capitalism; hence arises a necessary relation between the evolution of capitalist societies and the evolution of social configurations.

Marx argues that the basic social relationship within capitalism is a power relation between capital and labor. This power relation is expressed directly through a market mode of economic integration. Thus, the proportion of national product set aside for wages and profits (which includes rents and interest) is determined by the outcome of a class struggle between the representatives of labor (now usually the unions) and capital (usually the employers). Marx also argues that the power relation between the two great classes in society can be understood only in terms of the particular historical conditions achieved with the emergence of the capitalist order. Thus, labor power has to assume a commodity character, which means that it can be "freely" bought and sold in the market and that the laborer has legal rights over the disposition of his or her labor. Ownership and control over the means of production gives capital its power over labor, since the laborer has to work in order to live and the employer holds control over the means of work. A relatively stable power relation between capital and labor requires for its maintenance a wide variety of institutional, legal, coercive, and ideological supports, most of which are either provided or managed through state institutions.

The power relation between capital and labor may be regarded as the primary force of class structuration in capitalist society. However, this force does not necessarily generate a dichotomous class structure. The two-class model that Marx presents in volume 1 of *Capital* is an assumed relation through which he seeks to lay bare the exploitative character of capitalist production – it is not meant as a description of an actual class structure (*Capital* 1:167–70 and 508–10; 2:421). Marx also distinguishes between the roles of capital and labor and the personifications of those roles – the capitalist, although functioning as a mere personification of capital much of the time, is still a human being. The concepts of "class" and "class role" function in *Capital* are analytic constructs. Yet Marx often used the dichotomous model of class structure as if it had an empirical content, and in his more programmatic writings he insists that socialism will be achieved only through a class struggle that pits the capitalist class against the proletariat.

The reason for this stance is not hard to find. Marx attributed the exploitative character of capitalist society to the capital-labor relation, and he also traced the innumerable manifestations of alienation back to this one fundamental source. These negative aspects of capitalist society could be transcended in Marx's view only by transcending the power relation that permitted the domination of labor by capital. The analytic constructs of *Capital* consequently become normative (ought-to-be) constructs in his programmatic writings. And if actual class struggle crystallized around the capital-labor relation, then both the analytic and the normative constructs would come to take on an empirical validity as descriptions of actual social configurations.

But social configurations could crystallize along quite different lines in an actual situation. In *The Eighteenth Brumaire of Louis Bonaparte*, for example, Marx analyzes conflict in the France of 1848–51 in terms of the class interests of lumpenproletariat, industrial proletariat, petite bourgeoisie, industrialists, financiers, a landed aristocracy, and a peasantry. In using this more complex model of a social configuration, Marx was plainly not saying that France was not capitalist at that time. He was suggesting, rather, that capitalism had evolved at that particular time and in that particular place to a stage in which class interests (often of a myopic and nonrevolutionary sort) could and did crystallize around forces other than the fundamental power relation between capital and labor.

It is convenient to designate these forces as "secondary forces of class structuration" and to divide them into two groups. The first I shall call "residual," for they stem either from some historically prior mode of production or from the geographical contact between a dominant and a subordinate mode of production. In the early years of capitalism, residuals from the feudal order – a landed aristocracy and a peasantry, for example – were very important. Moreover, there is evidence that these residual features can be very persistent and last for centuries after the initial penetration of capitalist social relationships. The geographical expansion of capitalism into a global system has also created residuals. The patterns of dominance and subservience associated with colonialism and neocolonialism are products of an intersection between the forces of class structuration in a dominant capitalist society and forms of social differentiation in subordinate traditional societies. Residual elements may disappear with time or be so transformed that they become almost unrecognizable. But they can also persist. And insofar as transformed residuals become incorporated into the social structure of advanced capitalist societies, they help to explain the existence of transitional classes. Landlordism, preserved in a capitalist form, or a group subjected to neocolonial domination and transformed into a relatively permanent underclass (blacks, Puerto Ricans, and Chicanos in the United States, for example) are the kinds of features in a social configuration that have to be explained in terms of the residual forces of class structuration.

The other forces of class structuration derive from the dynamics of capitalist society. These "derivative forces," as I shall designate them, arise because of the necessities generated by the need to preserve the processes of capital accumulation through technological innovation and shifts in social organization, consumption, and the like. We can identify five such forces (following Giddens 1973), and we shall consider them briefly in turn.

The Division of Labor and Specialization of Function

The expansion of production requires improvements in labor productivity and in the forms of industrial organization, communication, exchange, and distribution. These improvements usually mean an increasing division of labor and specialization of function. As the technical and organizational basis of society changes, so there must be concomitant shifts in social relationships which create the potential for social differentiation. The distinction between manual and intellectual work, for example, may be reflected in the social distinction between blue-collar and white-collar workers. At the same time the growing complexity of economic organization may require the emergence of specialized financial intermediaries in the economy (banking and other financial institutions), which may be reflected in distinctions between financiers and industrialists within the capitalist class as a whole. The division of labor and specialization of function may fragment the proletariat and the capitalist class into distinctive strata. Social conflict may take place between strata and thus replace class struggle in the Marxian sense as the guiding principle of social differentiation.

Consumption Classes or Distributive Groupings

The progress of capitalist accumulation may be inhibited by the lack of an effective demand for its material products. If we leave aside the growth of demand inherent in demographic growth and tapping export markets, effective demand depends upon the creation of an internal market to absorb the increasing quantities of material products. Marx (*Grundrisse,* 401–23) argues that the creation of new modes of consumption and of new social wants and needs is essential to the survival of capitalism – otherwise capital accumulation faces an inpenetrable barrier of fixed demand, which means overproduction and crisis. Underconsumption, though not the fundamental underlying cause of capitalist crises (see Harvey 1982), is often a pervasive manifestation of crisis and, as such, has to be confronted directly as a key political and economic problem. Malthus (1951, 398–413), in first proposing a version of the Keynesian theory of effective demand, had argued that the existence of a class of "conspicuous consumers" (primarily the landed aristocracy in his time) was a necessity if sufficient effective demand were to be sustained to permit the accumulation of capital. Malthus's perspective is an interesting one. Not only does he suggest that specific mechanisms have to be employed to stimulate consumption, but that certain consumption classes have to exist to ensure sustained consumption. If this is the case, then social differentiation arises in the sphere of consumption. Distinctive consumption classes or distributive groupings are therefore bound to emerge in the course of capitalist history (Giddens 1973, 108–10). Since it is empirically observable that life-style and consumption habits vary across different strata in the population, and since this is an important differentiating feature in modern society, we may conclude that the emergence of distinctive consumption classes is inherent in the dynamics of capitalist society. Social differentiation can be structured, therefore, according to distribution and consumption criteria (*Grundrisse,* 402).

Authority Relations

The nonmarket institutions in society must be so ordered that they sustain the power relation between capital and labor and serve to organize production, circulation, and distribution. Marx (*Capital* 1:330–33) argues, for example, that cooperative activity in production requires a "directing authority" and that as capitalist production becomes more elaborate, so a specialized group of workers – administrators, managers, foremen, and the like – must assume an authority role in the direction of production. For the economy as a whole these management functions lie largely in the sphere of state activity – understood as the collective amalgam of legal, administrative, bureaucratic, military and political functions (Miliband 1969). Within this sphere, and within the corporate enterprise, authority relations are the basis for social relationships. In general, the structure of authority relations is coherent with the necessities imposed by the dynamics of accumulation within a social system organized along capitalist lines. But the authority relations appear independent of the relation between capital and labor and indeed are, to a certain degree, autonomous in their functioning (Poulantzas 1973). The structure of authority relations can, therefore, provide a basis for social differentiation within the population. Marx (*Theories of Surplus Value,* pt. 2:573) thus writes about the significance of "the constantly growing number of the middle classes [who] stand between the workman on the one hand and the capitalist and landlord on the other."

Class-Consciousness and Ideology

Marx argues that a class will become an observable aggregate of individuals only when that aggregate buries all the differences within it and becomes conscious of its class identity in the struggle between capital and labor. Since capitalism has evolved and survived, then presumably it has in part done so by an active intervention in those processes whereby class-consciousness in the Marxian sense is created. There is, as it were, a struggle for the mind of labor between, on the one hand, a political class-consciousness directed toward the transcendence of the capital-labor relation and, on the other hand, states of social awareness which allow of social differentiations consistent with the accumulation of capital and the perpetuation of the capital-labor relation. The struggle for the mind of labor is a political and ideological struggle. Marx considered that, in general, the ruling ideas in society are ever the ideas of the ruling class. Mass literacy and mass education have the effect of exposing the masses to a dominant bourgeois ideology, which seeks to produce states of consciousness consistent with the perpetuation of the capitalist order. Mass culture, or what Marcuse (1968) calls "affirmative culture," has the function of depoliticizing the masses rather than enlightening them as to the real source of alienation in society (see *Consciousness and the Urban Experience,* chap. 5).

Certain parallel processes can be observed in the political sphere. The survival of capitalism necessitates an increasing state interventionism, which, far from being neutral, actively sustains the power relation of capital over labor. In a given instance the state may throw its weight to the side of labor in order to restore some kind of balance between profits and wages, but state intervention is never geared to the transcendence of the capital-labor relation. Yet the state *appears* to be neutral. In part this appearance is real, for state institutions frequently arbitrate between factions of the ruling class (between financiers and industrialists, for example) and between strata of the working population. The separation between the economic and political-administrative spheres, which typically arises under capitalism, also permits the state to appear as a neutral party in economic conflict. At the same time the prospects for legal and political equality held out in the political sphere tend to divert attention from the inevitable subordination of labor to capital in the marketplace. This separation between economy and polity has been, as Giddens (1973, 202–7) points out, a fundamental mediating influence on the production of class-consciousness and social awareness in capitalist society. It typically feeds trade-union consciousness on the part of labor and a distinctive kind of middle-class awareness on the part of intermediate groups in the authority structure which focuses on civil and political liberties to the exclusion of questions of economic control.

Political and ideological struggles, and the manipulation of both, have great significance for understanding the states of consciousness of various strata within the population. Only in terms of such consciousness can we explain how and why a particular problem (say, unemployment) will elicit as a response conflict between capital and labor rather than conflict within labor. The second type of conflict might be between, say, the regularly employed and a largely unemployed underclass, that may also be a racial or ethnic minority. The first kind of conflict poses a threat to the capitalist order, whereas the latter kind of conflict does not. It is obviously in the interest of capitalism to transform conflict of the first sort into conflict of the latter variety. Consequently, bourgeois ideology and politics typically seek to forge a consciousness favorable to the perpetuation of the capitalist order and actively seek out ways to draw social distinctions along lines other than that between capital and labor. I take up these questions in detail in *Consciousness and the Urban Experience.*

Mobility Chances

The accelerating pace of change in the organization of production, exchange, communication, and consumption necessitates considerable adaptability in the population. Individuals must be prepared to alter their skills, geographical locations, consumption habits, and the like. This means that mobility chances must always be present within the population. Yet a completely open society as far as mobility is concerned would undoubtedly create considerable instability. In order to give social stability to a society in which social change is necessary, some systematic way has to be found for organizing mobility chances. This entails the structuring of mobility chances in certain important ways.

In capitalist society, mobility is organized so that most movement takes place between one stratum within the division of labor and another (from, say, the manual to the white-collar category). The mechanisms for achieving this controlled kind of mobility appear to lie in part in the differential distribution (both socially and geographically) of opportunities to acquire what Giddens (1973, 103) calls "market capacity" – that bundle of skills and attributes which permits individuals to market their labor power within certain occupation categories or to operate in certain functional roles. Restrictions and barriers to mobility chances give rise to social differentiations. Insofar as professional groups, for example, have better access to the acquisition of market capacity for their children, a professional "class" may become self-perpetuating. Once intergenerational mobility is limited, social distinctions become relatively fixed features in the social landscape and provide the possibility for the crystallization of social differentiation within the population as a whole.

The argument so far suggests that we can identify three kinds of forces making for social differentiation within the population:

1. A primary force arising out of the power relation between capital and labor
2. A variety of secondary forces arising out of the contradictory and evolutionary character of capitalism which encourage social differentiation along lines defined by (a) the division of labor and specialization of function, (b) consumption patterns and life-style, (c) authority relations, (d) manipulated projections of ideological and political consciousness, and (e) barriers to mobility chances
3. Residual forces reflecting the social relations established in a preceding or geographically separate but subordinate mode of production

In general we can see a perpetual struggle amongst these forces between those that create class configurations antagonistic to the perpetuation of the capitalist order and those that create social differentiations favorable to the replication of capitalist society.

II. RESIDENTIAL DIFFERENTIATION AND THE SOCIAL ORDER

The accumulation of capital on a progressively increasing scale has set in motion a distinctive and rapidly accelerating urbanization process. The distinctive features of this process need not delay us here (see Harvey 1973; Castells 1972; Lefebvre 1970, 1972). For the purpose at hand it is sufficient to note the progressive concentration of the population in large urban centers. There has been a parallel fragmentation of social structure as the primary, residual, and derivative forces of social differentiation have interacted over a century or more. Let us now locate these processes of progressive concentration and social fragmentation in the built environment we call the city and fashion some basic hypotheses to connect residential differentiation with social structure. Four hypotheses can be stated:

1. Residential differentiation is to be interpreted in terms of the reproduction of the social relations within capitalist society
2. Residential areas (neighborhoods, communities) provide distinctive milieus for social interaction from which individuals to a considerable degree derive their values, expectations, consumption habits, market capacities, and states of consciousness
3. The fragmentation of large concentrations of population into distinctive communities serves to fragment class-consciousness in the Marxian sense and thereby frustrates the transformation from capitalism to socialism through class struggle, but
4. Patterns of residential differentiation reflect and incorporate many of the contradictions in capitalist society; the processes creating and sustaining them are consequently the locus of instability and contradiction

These hypotheses, when fleshed out and if proven, provide a necessary link between residential differentiation and the social order. In the short space of this chapter, I can only sketch in a very general argument in support of them.

Residential differentiation in the capitalist city means differential access to the scarce resources required to acquire market capacity (Giddens 1973; Harvey 1973, chap. 2). For example, differential access to educational opportunity – understood in broad terms as those experiences derived from family, geographical neighborhood and community, classroom, and the mass media – facilitates the intergenerational transference of market capacity and typically leads to the restriction of mobility chances. Opportunities may be so structured that a white-collar labor force is reproduced in a white-collar neighborhood, a blue-collar labor force is reproduced in a blue-collar neighborhood, and so on. The community is the place of reproduction in which labor power suitable for the place of production is reproduced. This is a tendency only, of course, and there are many forces modifying or even offsetting it. And the relationships are by no means simple. Market capacity, defined in terms of the ability to undertake certain kinds of functions within the division of labor, comprises a whole set of attitudes, values, and expectations as well as distinctive skills. The relationship between function and the acquisition of market capacity can sometimes be quite tight – thus, miners are for the most part reproduced in mining communities. But in other cases the relationship may be much looser – a white-collar grouping, for example, comprises a wide range of occupational categories but is still differentiable, both socially and spatially, from other groupings.

Residential groupings that reproduce labor power to meet the needs of an existing division of labor may also form a distinctive grouping from the standpoint of consumption. Such a coalescence gives residential differentiation a much more homogeneous character. One thread of necessity as opposed to contingency in this relationship lies in the consumption of education, which unifies a consumption class with a grouping based in the division of labor. This thread is too slender to hang a proof on (although in the United States the connections between residential differentiation and the quality of education are very strong and a constant source of conflict and social tension). The full story rests on showing how attitudes generated out of the work experience imply certain parallel attitudes in the consumption sphere. To trace this connection is difficult, but it appears reasonable to suppose that the quality of the work experience and the attitudes necessary to perform that work under specific social conditions must be reflected, somehow or other, by attitudes and behaviors in the place of residence.

The relationships between values, consciousness, ideology, and life experiences are crucial – and they are the most profoundly difficult to unravel. From the standpoint of the creation of residential differentiation, it is plain that individuals do make choices and do express preferences. To sustain the argument, therefore, I have to show that the preferences and value systems, and perhaps even the choices themselves, are produced by forces external to the individual's will. The idea of an autonomously and spontaneously arising consumer sovereignty as the explanation of residential differentiation could be fairly easily disposed of (even though it is the prevalent myth that underlies conventional theories of residential differentiation). But it is more difficult to know what exactly to put in its place. And it is far too glib to attribute everything to the blandishments of the "ad-men," however important they may be.

If we ask, however, where peoples' values come from and what is it that creates them, then it is plain that the community provides a social milieu out of which distinctive value systems, aspirations, and expectations may be drawn. The neighborhood is, as it were, the primary source of socialization experiences (Newson and Newson 1970). Insofar as residential differentiation produces distinctive communities, we can expect a disaggregation of this process. Working-class neighborhoods, for example, typically produce individuals with values conducive to being in the working class; and these values, deeply embedded as they are in the cognitive, linguistic, and moral codes of the community, become an integral part of the conceptual equipment that individuals use to deal with the world (Giglioli 1972). The stability of such neighborhoods and of the value systems that characterize the people in them have been remarkable considering the dynamics of change in most capitalist cities. The reproduction of such value systems facilitates the reproduction of consumption classes as well as groupings with respect to the division of labor while it also functions to restrict mobility chances. Values and attitudes toward education, for example, vary greatly and affect the consumption of education – one of the main means of obtaining mobility chances (Robson

1969). The homogenization of life experiences which this restriction produces reinforces the tendency for relatively permanent social groupings to emerge within a relatively permanent structure of residential differentiation. Once this is translated into a social awareness that has the neighborhood or community as the focus, and once this form of social awareness becomes the basis for political action, then community-consciousness replaces class-consciousness (of the Marxian sort) as the springboard for action and the locus of social conflict.

Once such groupings form, it is relatively easy to understand how they may be perpetuated. But we also have to understand the history of such groupings because contemporary social differentiations have been arrived at by successive transformations and fragmentations of preceding social configurations. The reciprocity exhibited in working-class neighborhoods is to a large degree a defensive device constructed out of the transformation under capitalism of a well-tried and ancient mode of economic integration (Harvey 1973, chap. 6). In the United States, immigrant waves at particular periods in the evolution of the capitalist division of labor gave a strong ethnic flavor to certain occupational categories as well as to certain residential neighborhoods; both persist to the present day. The continued domination of blacks following the transformation of slavery, and the more modern neocolonial domination of Puerto Ricans and Chicanos, have produced the ghetto as a Third World colony in the heart of the American city, and it is broadly true that the underclass in American society is identified with neocolonial repression based in racism (Blaut 1974). The historical roots of social and residential differentiation are important. But then so also are the processes of social transformation that produce new social groupings within a social configuration.

Consider, for example, the emergence of a distinctive middle class, literate and skilled in mental labor, possessed of the dominant bourgeois ideology that McPherson (1962) felicitously calls "possessive individualism," attached as a consequence to certain distinctive modes of consumption, imbued with a political view that focuses on civil and political liberties, and instilled with the notion that economic advancement is solely a matter of individual ability, dedication, and personal ambition (as if everyone could become a successful doctor, lawyer, manager, and the like, if only they tried hard enough). The emergence of such a middle class over the past century or so has become etched into the city by the creation of distinctive middle-class neighborhoods with distinctive opportunities to acquire market capacity. In more recent times, affluent workers and white-collar employees have been encouraged to copy the middle-class life-style. And in the American city this process has been associated since the 1930s in particular with a strong suburbanization process. How do we explain the way in which the emergence of such social groupings relates to the process of residential differentiation?

The answer to this question depends in part upon an understanding of the processes whereby residential differentiation is produced by the organization of forces external to the individual or even to the collective will of the particular social grouping. These processes stretch back over a relatively long time period, and it is probably the case that residential differentiation in the contemporary sense was well established in most major cities in both the United States and Britain by 1850. In certain basic respects the processes have not changed, however, for we still have to turn to the examination of the activities of speculator-developers, speculator-landlords, and real estate brokers, backed by the power of financial and governmental institutions, for an explanation of how the built environment and residential neighborhoods are actually produced. I have attempted a full description of this process in the American city elsewhere (see Chap. 3), and so I shall merely offer a summary of it here.

Financial and governmental institutions are hierarchically ordered by authority relations broadly consistent with the support of the capitalist order. They function to coordinate "national needs" (understood in terms of the reproduction of capitalist society and the accumulation of capital) with local activities and decisions – in this manner, micro- and macro-aspects of housing market behavior are coordinated. These institutions regulate the dynamic of the urbanization process (usually in the interest of accumulation and economic crisis management) and also wield their influence in such a way that certain broad patterns in residential differentiation are produced. The creation of distinctive housing submarkets (largely through the mortgage market) improves the efficiency with which institutions can manage the urbanization process. But at the same time it limits the ability of individuals to make choices. Further, it creates a structure that individuals can potentially choose from but that they cannot influence the production of.

If residential differentiation is in large degree produced, then individuals have to adapt their preferences. The market mechanism curtails the range of choice (with the poorest having no choice, since they can take only what is left over after more affluent groups have chosen). The shaping of preferences of more affluent groups poses a more serious problem. The ad-man plays an important role, and considerations of status and prestige are likewise important. Consider, also, a white-collar worker forced to suburbanize (by a process I have elsewhere dubbed "blow out" – Harvey 1973, chap. 5) because of deteriorating conditions in the inner city; the preference in this case may be a somewhat shallow post-hoc rationalization of a "choice" that really was no choice. Dissatisfaction within such a group can easily surface. For example, a suburbanite angered by the prospect of gasoline shortages and recollecting the convenience of inner-city living complains that "we have all been had" because in order "to mold into the lifestyle dictated by builders, developers, and county planners, I have no choice but to provide my family with two automobiles" – one to get to work and the other to operate a household (*Baltimore Sun*, February 11, 1974). The consumption values attached to suburban living are plainly not open to choice once the location decision is made, and that decision may itself not be the outcome of a real choice.

Indeed, a strong argument can be made that suburbanization is a creation of the capitalist mode of production in very specific ways. First, suburbanization is actively produced because it sustains an effective demand for products and thereby facilitates the accumulation of capital. Second, the changing division of labor in capitalist society has created a distinctive group of white-collar workers who, largely by virtue of their literacy and their work conditions, are imbued with the ideology of competitive and possessive individualism, all of which appears uniquely appropriate for the production of a mode of consumption which we typically dub "suburban." It is intriguing to note that since the 1930s the United States has experienced the most sustained rate of economic growth (capital accumulation), the greatest growth in the white-collar sector, and the most rapid rate of suburbanization of all the advanced capitalist nations. These phenomena are not unconnected.

We can, thus, interpret the preference for suburban living as a created myth, arising out of possessive individualism, nurtured by the ad-man and forced by the logic of capitalist accumulation. But like all such myths, once established it takes on a certain autonomy out of which strong contradictions may emerge. The American suburb, formed as an economic and social response to problems internal to capitalist accumulation, now forms an entrenched barrier to social and economic change. The political power of the suburbs is used conservatively, to defend the life-style and the privileges and to exclude unwanted growth. In the process a deep irrationality emerges in the geography of the capitalist production system (residential and job opportunities may become spatially separated from each other, for example). The exclusion of further growth creates a further problem, for if a "no-growth" movement gathers momentum, then how can effective demand and capital accumulation be sustained? A phenomenon created to sustain the capitalist order can in the long run work to exacerbate its internal tensions.

This conclusion can possibly be extended to all aspects of social and residential differentiation. The social differentiations reproduced within the capitalist order are so structured as to facilitate the reproduction of the social relations of capitalism. As a result community-consciousness rather than class-consciousness in the Marxian sense is dominant in the capitalist city. In this fashion the danger of an emergent class-consciousness in the large concentrations of population to be found in urban areas has been averted by the fragmentation of class-consciousness through residential differentiation.

But community-consciousness with all of its parochialisms, once created, becomes deeply embedded, and it becomes just as difficult to piece it together in a configuration appropriate to the national interest (as perceived from the standpoint of capital accumulation) as it is to transform it into a class-consciousness antagonistic to the perpetuation of the capitalist order. In order, therefore, to maintain its own dynamic, capitalism is forced to disrupt and destroy what it initially created as part of its own strategy for self-preservation. Communities have to be disrupted by speculative activity, growth must occur, and whole residential neighborhoods must be transformed to meet the needs of capital accumulation. Herein lie both the contradictions and the potentials for social transformation in the urbanization sphere at this stage in our history.

Residential differentiation is produced, in its broad lineaments at least, by forces emanating from the capitalist production process, and it is not to be construed as the product of the autonomously and spontaneously arising preferences of people. Yet people are constantly searching to express themselves and to realize their potentialities in their day-to-day life-experiences in the workplace, the community, and the home. Much of the micro-variation in the urban fabric testifies to these ever-present impulses. But there is a scale of action at which the individual loses control of the social conditions of existence in the face of forces mobilized through the capitalist production process (in the community this means the congeries of interests represented by speculators, developers, financial institutions, big government, and the like). It is at this boundary that individuals come to sense their own helplessness in the face of forces that do not appear amenable, under given institutions, even to collective political mechanisms of control. As we cross this boundary, we move from a situation in which individuals can express their individuality and relate in human terms to each other to one in which individuals have no choice but to conform and in which social relations between people become replaced by market relations between things.

Residential differentiation on this latter scale plays a vital role in the perpetuation and reproduction of the alienating social relationships of capitalist society. Yet in the process of seeking stratagems for self-perpetuation, the forces of capitalist accumulation create value systems, consumption habits, states of awareness and political consciousness, and even whole built environments, which, with the passage of time, inhibit the expansion of the capitalist order. The permanently revolutionary character of capitalist society perpetually tears down the barriers it has erected to protect itself. The constant reshaping of urban environments and of the structures of residential differentiation are testimony to this never-ending process. Instead, therefore, of regarding residential differentiation as the passive product of a preference system based in social relationships, we have to see it as an integral mediating influence in the processes whereby class relationships and social differentiations are produced and sustained. It is clear, even at this preliminary stage in the analysis, that the theory of residential differentiation has much to offer as well as to gain from a thoroughgoing integration with general social theory.

6

The Place of Urban Politics in the Geography of Uneven Capitalist Development

There is an emerging consensus, as dangerous as it is unfounded, that the fluid movement of urban and regional politics cannot be incorporated into any rigorous statement of the Marxist theory of capital accumulation. The breadth of the consensus is quite surprising. It includes not only critics like Saunders (1981) who naturally tend to view any version of the Marxist theory with jaundiced eye but also a number of past sympathizers and current practitioners within the Marxist tradition. Mollenkopf (1983), for example, says quite firmly that an adequate theory of politics cannot be built upon Marxist propositions and that politics and government have to be viewed as "independent guiding forces" overriding economic considerations. More serious has been Castells's apparent defection from the Marxist fold. In *The City and the Grassroots* (1983, 296–300), he confesses that his "intellectual matrix" in the Marxist tradition "was of little help from the moment we entered the uncertain ground of urban social movements." The problem, he asserts, lies "deep in the core of the Marxist theory of social change," which has never overcome a duality between the logic of capital accumulation and historical processes of class struggle. Conceding the "intelligence" of Saunders's critique, Castells firmly rejects the idea that the city and space can be understood in terms of the logic of capital. He even doubts the relevance of class concepts and class struggle to understanding urban social movements. He seeks to build a more complicated reading of history, cities, and society out of "the glorious ruins of the Marxist tradition."

Defections of this magnitude testify to the depth of frustration which many feel as they try to bring the generalities of Marxian theory to bear on specific local and conjunctural events. In this chapter I try to confront this malaise directly. I try to show how, why, and within what limits a "relatively autonomous" urban politics can arise and how that "relative autonomy" is not only compatible with but also necessary to the processes of capital accumulation. In so doing, I accept that the problems posed by defectors and critics alike are real and not imagined. I also accept that the general theory of accumulation has not always been specified in a way that makes it easy to address urban processes and that there is a good deal of intellectual baggage (to say nothing of dogmatism) within the Marxist tradition which hinders rather than helps in the search for penetrating analyses and viable alternatives.

I cannot, in a single chapter, address all the problems that need to be addressed. I shall therefore rest content with the construction of a relatively simple line of argument which derives its strength, I think, from depicting accumulation as a spatiotemporal process at the very outset. I begin with the observation that the exchange of labor power is always spatially constrained. A fundamental defining attribute of an urban area is the geographical labor market within which daily substitutions of labor power against job opportunities are possible. After consideration of how capitalists act in the context of localized labor markets, I shall go on to show how unstable class alliances form within a loosely defined urban region. These alliances parallel the tendency for an urban economy to achieve what I call a "structured coherence" defined around a dominant technology of both production and consumption and a dominant set of class relations. The alliances, like the structured coherence they reflect, are unstable because competition, accumulation and technological change disrupt on the one hand what they tend to produce on the other. Here lies a political space within which a relatively autonomous urban politics can arise. That relative autonomy fits only too well into the geographical dynamics of accumulation and class struggle. In fact, it becomes a major means of bringing together the logic of capital accumulation with the history of class struggle. To the degree that different urban regions compete with each other – and Robert Goodman (1979) is not alone in regarding state and local governments as the "last entrepreneurs" – so they set the stage for the uneven geographical development of capitalism. Urban regions compete for employment, investment, new technologies, and the like by offering unique packages of physical and social infrastructures, qualities and quantities of labor power, input costs, life-styles, tax systems, environmental qualities, and the like. The effect of competition is, of course, to discipline any urban-based class alliance to common capitalist requirements.

But there is another side to all this. Capitalism is a continuously revolutionary mode of production. Speculative innovation in production processes is one of its hallmarks. But innovation in production requires parallel innovation in consumption. It also requires innovation in social and physical infrastructures, spatial forms, and broad social processes of reproduction. Innovation must extend to life-styles, organizational forms (political, cultural, and ideological as well as bureaucratic, commercial, and administrative) and spatial configurations. The ferment of urban politics and the diverse social movements contained therein is an important part of such an innovative process. It is as speculative, innovative, and unpredictable in its details as are capitalist processes of product innovation and technological and locational change. Furthermore, such behavior presupposes a wide range of individual liberties so that individuals, organizations, and social groups can probe in all manner of different directions. The more open the society, the more innovative it will likely be in all these different respects. But successful innovation and successful probing under capitalism mean profitable probing. Here the crude logic of capitalist rationality comes back into play. Cities, like entrepreneurs, can lose out to their competition, go bankrupt, or simply be left behind in the race for economic advantage. Urban politics then appear as the powerful and often innovative but in the end disciplining arm of uneven accumulation and uneven class struggle in geographic space. The discipline sparks conflict, of course, since the liberties conceded allow directions to be explored that are incompatible with or even antagonistic to capitalism. Cities may become hearths of revolution. But as the Paris Commune proved once and for all, there are then only two possible immediate outcomes. Either the revolution spreads and engulfs the whole of society, or the forces of reaction reoccupy the city and forcibly bring its politics under control.

I have long argued that urbanization should be understood as a process rather than as a thing. That process necessarily has no fixed spatial boundaries, though it is always being manifested within and across a particular space. When I speak of urban politics, then, it is in the broad sense of political processes at work within a fluidly defined but nevertheless explicit space. I do not mean the mayor and city council, though they are one important form of expression of urban politics. Nor do I necessarily refer to an exclusively defined urban region, because metropolitan regions overlap and interpenetrate when it comes to the important processes at work there. The urban space with which I propose to work is fixed only to the degree that the key processes I shall identify are confined within fixed spaces. To the degree that the processes are restlessly in motion, so the urban space is itself perpetually in flux.

I. THE URBAN LABOR MARKET

The working day, Marx long ago emphasized, is an important unit of analysis. It defines a normal time frame within which employers can seek to substitute one laborer for another and laborers can likewise seek to substitute one job opportunity for another. I therefore propose to view the "urban" in the first instance as a geographically contiguous labor market within which daily exchanges and substitutions of labor power are possible. Plainly, the geographical extent of that urban labor market depends upon the commuting range, itself historically determined by social and technological conditions. Furthermore, labor markets can overlap in space (fig. 12) and tend in any case to fade out over space rather than end at some discrete boundary. The urban labor market is better thought of as a complex drainage basin with firmer delineations at its center than on its periphery (Coing 1982). Nevertheless, a prima facie argument of considerable plausibility can be advanced which sees the urban-regional labor market as a unit of primary importance in the analysis of the accumulation of capital in space.

On a given day, the potential quantities and qualities of labor power available within this market area are fixed. This is a short-run inelasticity of supply to which capitalists may have to adapt. Whether or not they do so depends on a wide range of conditions, such as the gap between potential labor supply and that used (the labor surplus), the time horizon of capital investments, the ease of capital mobility, and the like. These are the questions I take up in section 2.

Urban labor markets exhibit all manner of peculiarities and imperfections. Labor power, Marx emphasized, is a peculiar commodity, unlike others in several important respects. To begin with, it is not produced under the control of capitalists but within a family or household unit. There also enters into the determination of its exchange value a whole host of moral, environmental, and political considerations. And finally, the use value of that labor power to the capitalist is hard to quantify exactly because of the fluidity of inherently creative labor processes. Storper and Walker (1984, 22–23) thus conclude that "labor differs fundamentally from real commodities because it is embodied in living, conscious human beings and because human activity (work) is the irreducible essence of social production and social life." The leveling and standardization achieved in other arenas of exchange can never be fully achieved in labor markets. Labor qualities always remain "idiosyncratic and place-bound." Each urban labor market, it then follows, is unique (Storper and Walker 1983, 1984).

The imperfections are likewise peculiar. Segmentations may exist in which certain kinds of jobs are reserved for certain kinds of workers (white males, women, racial minorities, recent immigrants, ethnic groups, etc.), while the geographical coherence may also be broken if certain kinds of workers (such as blacks in inner cities or women in the suburbs) are trapped by lack of transportation in geographically distinct submarkets or if different social groups have (by virtue of their incomes) differential access to different transportation modes. The principle of substitution is also modified by the nature of skill distributions in relation to the mix of labor processes within the urban region. There is, finally, the question of the available labor surplus

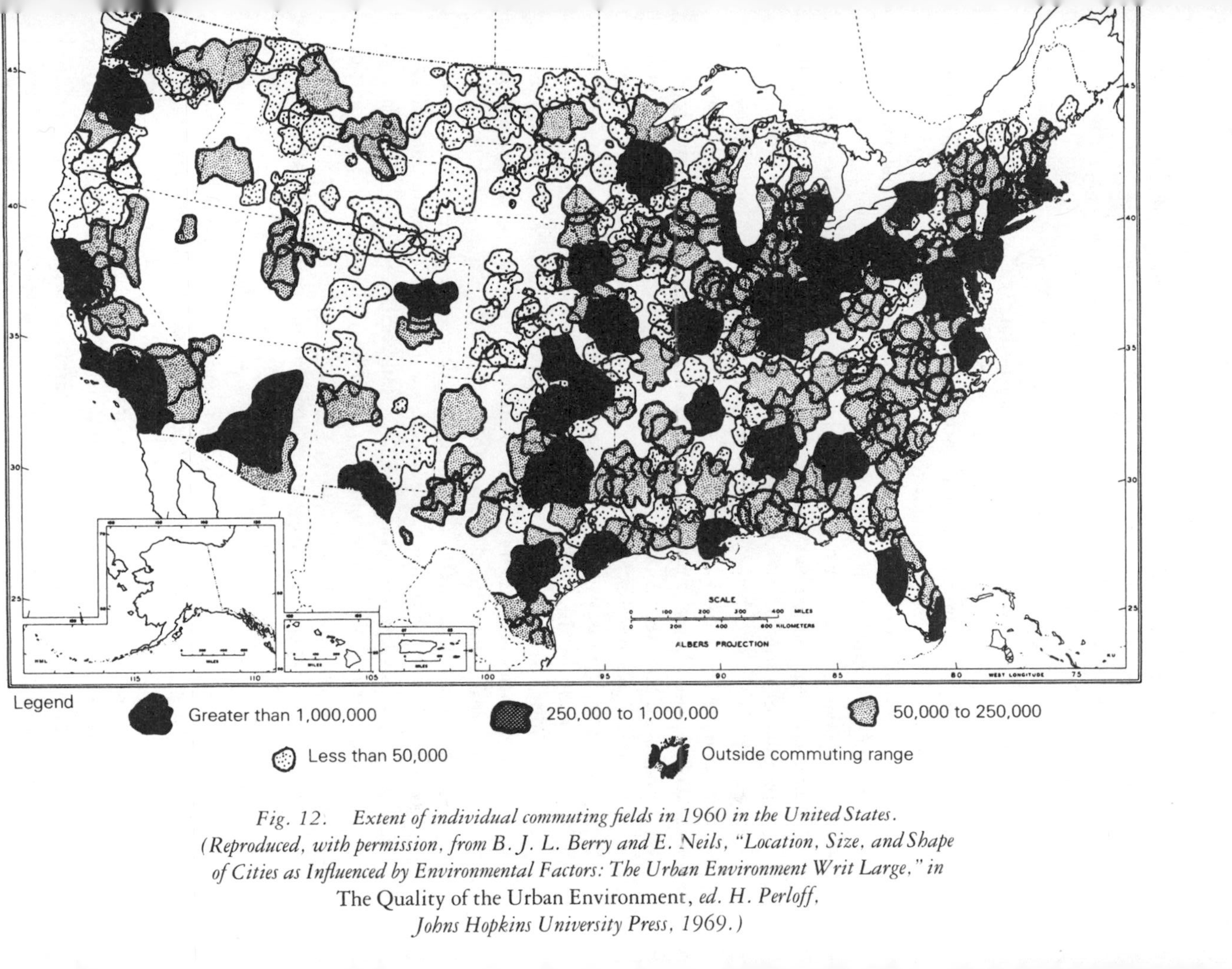

Fig. 12. Extent of individual commuting fields in 1960 in the United States. (Reproduced, with permission, from B. J. L. Berry and E. Neils, "Location, Size, and Shape of Cities as Influenced by Environmental Factors: The Urban Environment Writ Large," in The Quality of the Urban Environment, *ed. H. Perloff. Johns Hopkins University Press, 1969.)*

of unemployed, underemployed, or potentially employable people. Such reserves are usually stored in urban labor markets, though their mobilization sometimes poses peculiar problems. In all of these respects, the short-run rigidity of supply within the urban labor market to which capitalists must adapt is a *structured* rigidity.

The rigidities stand to be relaxed as we shift the time-horizon. Migration (occasional, periodic, permanent) releases the supply of labor power from the constraint of a daily commuting range. Supply is then contingent upon conditions in a "migration field" which, though usually to some degree geographically constrained, has the whole world as its possible outer limit. The supply of labor power within the urban labor market fluctuates on a weekly, monthly, or annual basis depending upon the balance of in- and out-migration. Minor adjustments of quantity and quality can quickly be achieved, but major flows pose more serious problems, again because of the peculiar qualities of labor power as a commodity. The bearer of that labor power, the living laborer, has to be fed, housed, and cared for somehow, and that means social costs and the provision of commodities to meet the laborer's daily needs. Even temporary or migrant workers require some level of provision, however dismal. Labor supply is harder to import than most other commodities and, once brought in, has a relatively permanent presence (except in the cases of special guest worker programs, temporary contract labor, and the like). Population growth also loosens long-run supply constraints, but again, adjustments are slow and hard to reverse. Shifts in labor qualities depend largely upon long-term processes of education (formal and informal) and cultural change (for example, the acquisition of appropriate work habits, the internalization of a "Protestant ethic," etc.). Internal adjustments in patterns of segmentation (the tightening or relaxation of discriminatory barriers between the races and sexes), together with changing social pressures toward the mobilization of latent labor reserves (the participation of women in the work force, the imposition of workfare provisions by the state, etc.), can also give greater flexibility to urban labor markets. In the long run, therefore, the supply of both quantities and qualities of labor power stands to be reasonably elastic, though constrained by social costs, long time-horizons for certain kinds of adjustment, and important irreversibilities.

Given such a condition, how, then, does an urban labor market work? First, consider matters solely from the point of view of capital accumulation. This means accelerating either quantitative expansion in the work force or qualitative change in response to organizational and technological revolutions in methods of production. How are the surpluses of labor power or untapped reserves of labor qualities (i.e., appropriate attitudes and particular skills) produced and maintained? Marx's "general law of capitalist accumulation" provides one kind of answer. The paths of accumulation and technological change intersect to produce a labor surplus (an industrial reserve army), no matter what the pace of population growth or in-migration. Technologically induced unemployment allows capitalists to operate on the supply as well as the demand end of the labor market, thus controlling wage levels, unemployment, and the like (Harvey 1982, chap. 6). The condition of the labor reserve then depends upon traditions of welfare arising out of socioeconomic and political evolution. The ability to switch into new technologies and organizational forms, however, depends on the adaptability of labor qualities. While the "general law" is useful as a first approximation, it requires much more elaboration if we are to understand the dynamics of urban labor markets.

Now consider the importance of economies of scale in the geographical space of an urban labor market. Those capitalists who operate in a line of production with low barriers to entry or easy capacity for expansion (which implies no large discrete increments in employment numbers due to "lumpiness" of other investments) are drawn to large urban labor markets because a small relative labor surplus in a large labor pool provides them individually with abundant reserves for start-up and expansion. Daily rigidities of labor supply can be more easily by-passed there. The counter-effect, however, is to draw such capitalists to agglomerate in large urban labor markets and so to push against the ceiling of total labor supply. In the absence of strong in-migration or population growth, this pushes up wage rates and tends to stimulate labor-saving innovations. And that implies a shift in demand for particular labor qualities. But a large urban labor market also gives individual capitalists abundant opportunities for substitution, thus reemphasizing the advantages of agglomeration. While the need for particular skills may be diminished for all but a small elite within the labor force, other qualities such as discipline, work attitudes, respect for authority, loyalty, and cooperation may become even more important and just as difficult to ensure. De-skilling in the narrow sense does not mean, as some suppose, that the question of labor qualities evaporates. Flexibility of labor qualities is, therefore, an important attribute of urban labor markets. The broad sense of "qualities" I am here invoking connects those attributes to the sociopsychological and cultural evolution of the labor force. While industries with large production units and lumpiness of investment lie outside these imperatives to some degree, the same tendency toward geographical concentration is given, this time within the form of industrial organization itself, while the long-term investments involved put a premium on stable quantities and qualities of labor supply (Storper and Walker 1983, 1984).

Now consider matters from the standpoint of the laborers. Marx relies in theory, as the capitalist does in practice, upon the "laborers' instincts for self-preservation" to ensure the daily and long-term reproduction of labor power.

Behind that blanket phrase lie complex processes of household formation, gender, family and kinship relations, personal networks, community solidarity, individual ambition, and the like. Such conditions are outside the direct control of the capitalists (or others), though there are innumerable indirect paths (religion, education, state programs, etc.) through which capitalists can influence them. The practices of labor reproduction are diverse, and the particular mix arrived at has profound implications for the qualities and quantities of labor supply within an urban region.

The "instincts for self-preservation" can take on individual or collective manifestations and are, in any case, open to a variety of social or psychological interpretations. While some workers may view individual migration as a means to personal economic advancement, others may choose to stay in place, organize, and fight collectively for improvements within the urban region. In between lies a range of other possibilities, perhaps best captured by the concept of distinctive working-class cultures that combine elements of individualism with certain habits of collectivism and common attitudes toward work, living, consumption, and "social progress." Furthermore, rising wage rates and expanding job opportunities are not the only motivations for migration. The psychosocial motivation to move to a large urban labor market may be just as important. The peculiar mix of freedoms and alienations, of hopes and risks which a large urban labor market offers are powerful incentives. Though the streets of Birmingham or Chicago may not be paved with gold, there is enough gold in circulation there that any one individual might reasonably hope for a piece of it. In the same way that capitalists are attracted to large urban labor markets out of economies of scale, so individual laborers are drawn for analogous reasons; the range of choice and possibilities for substitution are that much greater. The effect, however, is that individual laborers, acting in their own self-interest, tend to produce a surplus of labor power in a particular urban labor market and so undermine their own class interest. To the degree that individual capitalists engage in the same type of behavior, so considerable pressure builds toward increasing the scale of economic activity within an urban region. Nevertheless, pressures also exist for class-conscious labor organizations to seek ways to limit in-migration in order to preserve wage levels and conditions of working and living. They may do so by monopolizing certain kinds of jobs or simply by giving an unfriendly reception to new immigrants. Here, too, is a condition where differences of race, religion, ethnicity, and gender can be used as means of control or of selective integration of new labor supplies into an existing urban labor market. I make that point in order to show that labor market segmentations arise as much out of the laborers' desire to control the supply of labor power as they do out of employers' search to divide and rule.

The question of labor qualities deserves closer scrutiny. How do the laborers' instincts for self-preservation affect these? We are dealing here, as Storper and Walker (1984) emphasize, not only with the particular mix of skills and capacities to cope with this or that labor process but also with the whole sociology and psychology of work. Work habits, respect for authority, attitudes toward others, initiative and individualism, pride, and loyalty are some of the qualities that affect the productivity of labor power as well as its capacity to engage in struggle against capitalist domination. These qualities are not uniformly present within an urban labor market and are often highly differentiated, tending to produce their own kinds of structured rigidities. The product of a long history of sociopolitical development, they are often relatively inelastic in the medium and perhaps even in the long term. The qualities tend to be modified, as does their distribution within a given population by complex interactions of class-bound processes. Laborers can educate themselves, raise their own consciousness, struggle to bring whatever benefits they acquire to others through active political militancy, or try to monopolize certain skills, thus segmenting qualities in ways advantageous to subgroups (such as an aristocracy of labor, a layer of foreman-managers, or a particular job-skill structure). The motivations and ambitions of individual laborers, taken together with their collective endeavors, play a crucial role in the transformation of labor qualities independent of the capitalists' own motivations. The drive to self- and collective improvement is something to which the capitalist must often adapt, rather than something that the capitalist deliberately sets in motion.

Capitalists can also launch or support programs for the modification of labor qualities in ways that appear amenable to them. Other groups within the bourgeoisie (religious, educational, bureaucratic) have their own agendas and can enter into strategic alliances with either capitalists or laborers to press home private or state efforts to improve labor skills and qualities. The evolution of labor skills does not proceed, therefore, in ways narrowly functional for employers. Of course, evolutions in the labor process set a context within which such drives unfold and in some cases dictate the general paths down which the evolution of labor qualities must march. They can likewise create conditions that force laborers to adjust and adapt their own capacities and attitudes. But there is always an interaction, mediated by the laborers' instincts for self-preservation and advancement.

Labor qualities, once acquired, do not, unlike many other forms of investment, necessarily run down over time. The productivity of labor (like that of the soil, to use an analogy that Marx invokes to great effect) can build up over time, provided proper care is taken. The effect is to make each labor market even more unique, because the long processes of sociopolitical development within an urban region can build up unique mixes of qualities. The plundering of those qualities through de-skilling, overworking, bad labor relations, unemployment, and so forth can, however, lead, like soil mining, to the rapid depletion of a prime productive force. Attitudes of cooperation can be turned overnight into attitudes of violent confrontation; technical skills can be by-passed and lost; attitudes toward race and gender can be turned around to flare into conflicts within the labor market as well as within the workplace. The problem, of course, is that the coercive laws of competition tend to force individual capitalists into strategies of plundering, even when that undermines their own class interest. Whether or not such a result comes to pass depends upon the internal conditions of labor demand and supply, the possibilities of replenishment of labor reserves through migration either of labor power or of capital, and the capacity of capitalists to put a floor under their own competition by agreeing to some kind of regulation of the labor market.

But once the surpluses of labor power within an urban labor market are exhausted, capitalists have no options except to move elsewhere, seek out labor-saving innovations in order to create an industrial reserve where there was none before, force changes in the conditions of labor power utilization (reduce the social and legal restraints on the length of the working day, raise the rate of exploitation of women, children, and the aged), or simply mobilize the wholesale importation of labor surpluses from elsewhere. In practice, capitalists (or at least a faction of them) engage in all of these strategies well before the point of total absorption of the labor surplus is reached. Even minor increases in labor costs or minor threats of labor organization can lead to major switches in individual strategies of accumulation. But those strategies spark resistance and struggle, since by definition they threaten the security of labor.

The value of labor power (understood as the physical standard of living of labor) and of other important magnitudes in social life, such as the length of the working day, have to be viewed as an outcome not only of class-bound processes but also of active class struggle. The evidence for that is so overwhelming that it scarcely needs demonstration. What does need elucidation is the way in which those processes operate within the confines of geographically specific labor markets so as to emphasize rather than diminish the unique qualities of each. There are plenty of nineteenth-century studies, of which Engels's *Condition of the Working Class in England in 1844* (1971) is one of the most outstanding examples, which depict labor struggles as unfolding in specific urban and regional contexts. John Foster's (1974) comparative study of class structure, struggle, and labor process evolution in Oldham, Northampton, and South Shields is an interesting example of this genre of labor history. Divergence between urban labor markets during this period appears just as important as any trend toward uniformity. Yet even at this period we can spot a problem. The state is a unit of regulation which casts its net over a far vaster space than that of the urban labor market. Though the pressure for state regulation may arise within particular labor markets, the effect is to impose a surface veneer of uniformity across all labor markets. Class struggle through the nation state tends to downgrade urban labor markets to acceptable variations around some national norm. The differences between nations become greater than those within nations.

There is considerable evidence, therefore, that the geography of labor markets has evolved in the twentieth century into a more coherent hierarchy of international, national, regional, and urban labor markets. While the last cannot be understood without the others, I want to make the argument for the urban labor market as a fundamental unit of analysis within the hierarchy for exactly the same reason as I see it as a fundamental arena of class struggle and labor-force evolution. The political processes that transmit demands from urban to state levels and back again are, however, rather complex. State regulation may be concentrated in a few sectors and so have differential rather than uniform impacts upon urban labor markets. Enforcement can also vary from one place to another depending upon class consciousness and mobilization and the pressure among capitalists to circumvent the law. And to the degree that the state apparatus is itself decentralized, as in the United States, much regulation of labor markets dissolves into a mosaic of regional and even local differentiations. Collective bargaining at the national level likewise always leaves room for local variation, while unionization also varies from place to place. The strong local implantation of labor unionism still makes the title of "union town" meaningful. In the United States, for example, antiunion manuals advise keeping plants small (fewer than one hundred employees) and at least two hundred miles apart.

But there is much more to class struggle and the evolution of labor power than legal regulation and collective bargaining. All manner of informal arrangements can be arrived at, a whole culture of work, struggle, cooperation and social interaction evolved, which gives a unique coloration to labor qualities within an urban region. Religion, education, tradition, individual motivations, and patterns of collective mobilization integrate with the laborer's instincts for survival to produce a mosaic of urban labor markets which, though they may overlap and interpenetrate and integrate upward into regional and national configurations, form important units of analysis, if only because they remain the basic frame within which the working day finds its geographical range of possibilities. I shall therefore hold the "focus of resolution" at this level for purposes of further analysis, though with the clear recognition that it is not the only relevant geographical scale for looking at labor market behaviors.

II. SPACE, TECHNOLOGY, AND CAPITALIST COMPETITION IN URBAN LABOR MARKETS

The entrepreneurial search for excess profit is fundamental within the social relations of capitalism. Excess profits can be had by virtue of superior technology and organization or by occupying superior locations. In itself, of course, location means nothing: the capitalist needs privileged access (measured in time and cost) to raw materials, to intermediate products, services, and infrastructures (physical and social), final markets, and, of course, to labor supplies of the requisite quantity and quality.

The coercive laws of competition force capitalists to search out superior technologies and locations. This imparts strong technological and geographical dynamism to production, exchange, and consumption. Strong checks exist, however, to an unlimited dynamism. Change of technology entails costs, as does locational shift. A rational capitalist should not change either unless the excess profits outweigh the costs. But competition does not necessarily promote such rationality. In the same way that individual laborers may see the streets of every city paved with gold, so individual capitalists often see gold shimmering out of every technological gimmick and every new locational niche. Lured on, they may take action with disastrous consequences not only for themselves but also for the whole capitalist class. The search for excess profit means speculation, and speculation too easily breeds excess. In the face of that experience, a countervailing capitalist rationality can arise. Capitalists can seek to convert control over technology or location into monopoly privileges that put themselves outside of rather than merely ahead of their competition. Labor processes can be protected by industrial secrecy, patent laws, and the like. In some industries, such protections are weak (anyone can set up a sweatshop just by buying a few sewing machines and hiring labor), whereas in others (pharmaceuticals and electronics) the protections are usually stronger. The capacity to convert the "natural monopoly" of spatial location into business monopoly also varies from the strong protections inherent in the provision of utilities to the weak protections afforded builders and contractors.

Most capitalists would probably prefer to be outside of rather than merely ahead of their competition. The search for excess profit therefore divides into two streams: first, a competitive path searching out new technologies and locations for temporary gain; second, the search for monopoly power through exclusive command of technological or locational advantage. The drive to convert temporary into permanent monopoly advantage is more important than is generally realized. It is of particular relevance in explaining capitalist behavior in urban labor markets. Once favorably positioned, they may push to consolidate monopoly powers, move to prevent the infiltration of competition, try to seal in access to special qualities of labor supplies, lock up flows of inputs by exclusive subcontracting, and monopolize market outlets by equally exclusive franchises, dealerships, and the like. In so doing, they necessarily become deeply embroiled in the totality of political-economic processes operating within a particular labor market. They try, in effect, to manage their own positive externality effects and to capture the benefits of the urban synergism which they consciously help to promote.

This dual path to excess profit splits into another important duality to the degree that trade-offs exist between technological and locational advantages. A superior technology can compensate for an inferior location, and the opposite is also true. Under competition, some rough equilibrium should arise within the landscape of capitalist production in which trade-offs between technological and locational advantages would be nicely balanced for all producers. It would be a highly unstable landscape. The range of spatial competition is technologically defined (by economies of scale and the range of a good) at the same time as the appropriate technology depends upon the size and scale of the market area for both inputs and outputs (to say nothing of the fluid movement of labor supply). The search for excess profits through technological change is not, therefore, independent of the search for excess profits from location. Even in the absence of excessive speculation, competition simultaneously provokes shifts in spatial configurations and technological mixes, making "spatial equilibrium" (in the sense of classical location theory) an impossibility. The closer a space-economy approaches an equilibrium condition, the more the accumulation of capital and the search for excess profits will disrupt it (see Chap. 2).

But the pace of change of either sort is held back by the turnover time of the capitals engaged. Different kinds of goods require different working periods (varying from the daily baking of bread through the annual productions of agriculture, to the even longer times taken to produce, say, a hydroelectric dam or a power station). Furthermore, the different inputs can be turned over at different rates (machinery, buildings, etc.). The longer these working periods and turnover times (see Harvey 1982, chap. 8), the greater the geographical and technological inertia. New technologies and locations cannot be achieved until after the value embodied in the fixed capital employed has been fully recovered; unless, that is, a portion of the value is devalued, written off before its economic lifetime is out.

The longer these lifetimes, the more vulnerable such production systems become. Any withdrawal of the daily flow of labor (through strikes, outmigration, or transfer to other job opportunities within the urban labor market) spells danger for capitalists employing large quantities of fixed capital. Any introduction of new methods, new product lines, or new input configurations also puts existing systems of production and the fixed capital they employ in jeopardy. These problems are so serious that capitalists will not undertake long-term investments without the assurance of some stability in labor markets and without protection against excessive speculative innovation. Under such conditions, monopoly control of technology and location appears necessary, a vital means to guarantee conditions for long-term investment. But the trade-off between location and technology also enters in. Highly competitive firms in the realm of technology may carve up the world into monopolistically controlled spaces, while firms that are highly competitive in local markets (like building contractors) may be weakly competitive when it comes to technology. We thus end up with a four-way classification of monopolistic or competitive styles of seeking excess profits from technology or location. In practice, most firms probably end up closer to the center than to the peripheries of that frame.

The greater the monopoly power, the more the geographical landscape of capitalist production tends toward a relatively stationary state. The rhythms of technological and locational change are slowed down even beyond the point required to guarantee the proper amortization of fixed capital. The trouble, of course, is that such fixity is inconsistent with the further accumulation of capital. It may even undermine the capacity to wage class struggle against labor effectively. Herein lies a basis for stagnation and then crisis formation. A sudden break with past technological mixes and spatial configurations often entails massive devaluations of the preexisting capital and the breaking of powerful monopoly privileges, including those built up by labor organizations. The crisis "liberates" capital from the stagnation of its quasi-monopolistic chains; it permits new technologies to be deployed and new spatial configurations to be created. It also implies new patterns of labor bargaining and of social relations in production. The trade-offs between technology and location are radically disturbed, and new interurban, interregional, and international divisions of labor arise, only to embody new monopolistic elements. The geographical and technological landscape of capitalism is torn between a stable but stagnant calm incompatible with accumulation and disruptive processes of devaluation and "creative destruction."

When we look at these processes from the standpoint of the social, spatial, and technological divisions of labor, however, we note another set of restraints. The complementarity of many production flows and the "roundaboutness" of production techniques make it hard for firms to shift technique or location without parallel moves on the part of other firms. Considerable economies of scale attach to agglomeration in areas where a wide range of substitution (of inputs or of markets) is possible. The effect is to tend to confine innovation possibilities to restricted locations for competitive firms – hence the significance of large metropolitan areas (Chinitz 1958). Only in relatively monopolistic sectors can innovation disperse (to rural R & D establishments, for example). Shifts of location are likewise constrained by the difficulty of ensuring simultaneous moves by suppliers or market outlets. Only when firms have sufficient monopoly power can they drag their own suppliers and market outlets with them or economize on the use of transportation systems so that location no longer matters. The dynamism I have depicted in the case of individual competitive firms is therefore modified by conditions of complementarity within an overall spatial division of labor.

Capitalist behavior is thus ambiguous in relation to spatially defined urban labor markets. On the one hand, the thrust to gain monopoly privileges that put them above their competition, coupled with economies of agglomeration within a spatial division of labor, can lead firms to be both covetous and solicitous of tapping into and preserving the special privileges of exclusive access to labor supplies of a certain quantity and quality. They may even take active steps to help preserve labor qualities, compromise with labor demands in return for labor cooperation, and put some of their own resources behind the drive to enhance labor qualities (though always, of course, with an eye to those qualities they regard as advantageous to themselves). On the other hand, competition (either spatial or technological) can push them to ride roughshod from one type of labor market (social or spatial) to another with scant regard for the consequences of plundering the qualities and quantities of labor power. They are then forced to visit the costs of creative destruction on the labor force within the urban region with all manner of consequences, including the eruption of labor discontent and the triggering of awesome class struggle.

To summarize this section: The search for excess profits based on technological and locational advantage is limited by the monopolistic element of both. The trade-off between technology and location is an active factor within the formation of the geographical landscape of production and becomes more so to the degree that monopoly privileges attach to both within a spatial and technological division of labor. The process of accumulation requires, however, that monopoly privileges be broken. And that can be done only through processes of creative destruction, which means the devaluation of both capital and labor power. The capitalist landscape of production therefore lurches between the stabilizing stagnation of monopoly controls and the disruptive dynamism of competitive growth.

III. THE TENDENCY TOWARD "STRUCTURED COHERENCE" IN THE ECONOMY OF URBAN REGIONS

The class relation between capital and labor tends, under the conditions described, to produce a "structured coherence" of the economy of an urban region. At the heart of that coherence lies a particular technological mix – understood not simply as hardware but also as organizational forms – and a dominant set of social relations. Together these define models of consumption as well as of the labor process. The coherence embraces the standard of living, the qualities and style of life, work satisfactions (or lack thereof), social hierarchies (authority structures in the workplace, status systems of consumption), and a whole set of sociological and psychological attitudes toward working, living, enjoying, entertaining, and the like. We shall later see how the coherence also spawns a distinctive urban politics. How is this tendency toward structured coherence produced?

Laborers, we saw, are free – depending on their skills, the degree of competition between them, the forms of segmentation, and the levels of demand – to substitute job opportunities daily within a socially given commuting range. Wage rates and working conditions within sectors should therefore be more finely adjusted within than between urban labor markets. To reproduce labor power on a daily basis, workers must also spend much of what they earn on goods and services retailed within a similar socially given commuting range. Within that range, they are free to patronize this or that establishment, purchase this or that service, bid for this or that house, and press for this or that level of social provision. In so doing, they convert their wage into a certain style, standard, and pattern of consumption within a system of market areas for goods and geographically defined labor markets. Though the two market areas may not be coterminous for any one individual or family (those on the fringes in particular may work in one market area and shop in another), the effect is to tend to define an urban area as a unity of job and consumption opportunities. To be sure, the skills and status system and the various labor market segmentations prevent any equalization of wage rates, working conditions, living standards, or social provision within that urban region, while equally vigorous discrimination on the consumption side (racial, sexual, or religious discrimination in housing markets, in social provision, etc.) creates another layer of barriers to the equalization of living standards and life-styles. There is, nevertheless, a certain merit to considering an underlying unity of job and consumption opportunities as a norm around which such deviations pivot. This norm fixes, for example, a standard against which the sense of relative deprivation may be measured.

The conception to which this leads is of a daily exchange of labor power and a daily reproduction of labor power caught within the confines of some loosely defined field of commuting possibilities. From a purely technical standpoint this positions labor as an appendage of the circulation of capital within the urban region. Wage revenues circulate out of production only to enter back into production as a living laborer, fed, housed, rested, and ready for work. This "company store" image does not imply powerlessness on the part of the laborer, though it does limit the exercise of that power, short of revolution. Collective struggles within the confines of this appendage relation frequently affect the exact form of structured coherence achieved. This applies as much to the public provision of services (hence the significance of local politics) as it does to struggles over local wage rates, working conditions, and the nature, price, and quality of consumption goods and services. Within the frame I have defined, therefore, there is abundant oportunity for laborers to pursue the enhancement of skills, the construction of class organizations for mutual support (e.g., savings associations, mutual benefit societies), and the building of a basis for political power. Employers are likewise caught in the same dependency relation. They are as deeply beholden to the daily exchange of labor power for their profit as laborers are beholden to them for daily sustenance. Ownership and control over the means, mechanisms, and forms of production gives them key advantages in class struggle, but they can never escape the dependency relation, which, in the first instance, is necessarily articulated through daily labor markets.

Consider the broad contours of the struggle between capital and labor within the confines of the urban region. Technology offers capitalists a vital means of control which mediates the production of labor surpluses and wage rates and ranges as well as job structures and hierarchies. Struggles over the deployment of new technologies are fundamental to the kind of structured coherence achieved within the urban region. Struggles waged by either capitalists or laborers to sustain and enhance labor quantities and qualities are likewise important, since they define the range of technological possibilities. Cooperation, cooptation, and consent are also a part of class struggle. They have the advantage, for capitalist and laborer alike, of giving a certain measure of stability and security to both work and the standard of living, albeit under the overall domination of capital. Accommodation plays a vital role in giving a relatively stable structured coherence to production and consumption within the urban region.

Capitalists, however, can choose to buy and sell commodities on a daily basis within or without the urban region, depending, of course, on the nature of the good, transport costs, effective demand, relative prices, and the like. Commodity and service production within an urban region roughly divides (as economic base theory long ago argued) into locally produced and consumed goods and services and an "export" trade counterbalanced by "imports" of goods and services from other urban regions. The volume and qualities of the internal market are, however, important parameters for much of the capital operating there. Effective demand in that market depends upon wages paid out, new investments made, and revenues received (rents, interest, taxes, profits). Distribution relations therefore affect the kind of structured coherence achieved (the balance of luxury versus wage goods, between final consumption and new investment, for example). Struggles over distribution and over the forms of consumption unfold unevenly from one urban region to another and likewise contribute to the uniqueness of each (cf. Katznelson 1981).

Consider, now, the complexities that arise through spatial competition and complementarity between urban regions within a geographical division of labor. Low wage costs in a particular industry may make it easier to compete in other markets as well as internally. But low wages mean less local effective demand, which may reduce economies of scale for locally oriented production and thus undermine the capacity to compete within a geographical division of labor. I use tht example to illustrate the idea that high wage costs do not always undermine competitiveness but can sometimes improve it, depending on the sector. Processes of this type underline the unique position of each urban region in the kind of structured coherence achieved in relation to its position within the geographical division of labor. The principle I am working toward is this: that class struggles over wage rates, working conditions, consumption (public or private), distribution relations, and so forth within an urban region intersect with export-import relations within a geographical division of labor in highly specific ways. Determinations are reciprocal and, as we shall see, necessarily fluid and dynamic. The tendency toward structured coherence takes shape and is shaped by these reciprocal relations. The effect is to emphasize the uniqueness of geographical position as well as of the qualities of each urban region.

Within this broad idea there are two issues I want to take up for closer inspection. Consider, first, how the technological mix within an urban region tends to define production and consumption processes simultaneously. There are, to begin with, certain important sectors (transportation being one) where an industry serves both production and consumption simultaneously, automatically unifying the technology deployed in both. In other cases, the push to serve both producer and consumer markets arises out of the quest for economies of scale in the market (this is very much the case with electronics). Then there are less tangible but no less important unifying forces. Designers and workers familiar with a technology (its functions, maintenance, etc.) in the workplace can quickly adapt it to uses in the living space. But the inverse relation is also true. The kinds of skills children pick up from the technology of play form a basis for the skills they can bring to production: computer games are an important educational device. Indeed, pressure for innovation can just as easily begin in the home as in the workplace (the demand for labor-saving household gadgets, innovative games, etc.). The motivations and logic of development are, however, quite different. Competition, class struggle, and the need to coordinate production push capitalists toward innovation even in the face of the barrier of amortization of past investments.

In the consumption sphere, premature and planned obsolescence has to be produced by the mobilization of fashion, style, status-seeking, possessive individualism, or appeals to social progress. The effect, however, is to accentuate the parallel evolutions of the technologies of production and consumption within the urban region. Only in "export enclave" economies of some Third World cities do we find a strong separation in the technologies of the two spheres. The tendency in the urban economies of advanced capitalist economies is to forge powerful and significant links between the technologies of the two spheres. Indeed, a case can be made that the stronger the link, the more dynamic the urban economy will become.

Consider, second, the links between social relations in the workplace and in the living space. Here, too, the tendency toward structured coherence rests on parallel evolutions, though again under very different circumstances and with quite different motivations. Patriarchal relations within the family, for example, can be taken over wholesale in the organization of social relations in the workplace. Modifications in the workplace – themselves influenced by patterns of technological change and of labor demand (the need to mobilize women as part of an industrial reserve army, for example) – have implications for social relations in the household (cf. *Capital* 1:490). Ties of family and kinship and relations of gender and age adapt to new forms of industrial organization (cf. Hareven 1982) at the same time as employers are drawn willy-nilly to use those familial and familiar relations as means of control and cooptation. These qualities of labor power evolve jointly through the daily experience of living and working.

But I speak only of the *tendency* toward structured coherence because it exists in the midst of a maelstrom of forces that tend to undermine and disrupt it. Competition over technological change, product innovation, and social organization; class struggles over distribution; social relations of production and reproduction; shifting space relations; and the push to accelerate turnover times and accumulation all make for constant imbalances. Equilibrium could be achieved only by accident, and then only momentarily. To the degree that capitalism internalizes powerful contradictions – between growth and technological progress, between the growth of productive forces and the dominant social relations of production – so the economy of any urban region is always potentially crisis-prone. Overaccumulation and devaluation are perpetual threats that have to be contained.

The potential responses to that threat are very different and take the urban economy in quite different directions. First, increasing monopolization provides a way to control the disequilibrating processes. We have already seen how private monopolization (of space, technology, or both) is one effective solution to excessive and destructive competition, and we have also argued that the thrust toward monopolization within an urban economy is the first level at which the monopolization due to space may be procured. We now encounter a collective version of that solution. Vested interests in the status quo actively cooperate to contain the forces of disruption. And in so doing they reinforce, perhaps even try to institutionalize, the kind of structured coherence already achieved within the economy of the urban region. The pursuit of such a strategy has two disadvantages. It leads to both internal stagnation and loss of external competitiveness within the geography of accumulation.

The second direction seeks to resolve the contradictions through temporal and spatial displacement (cf. Chaps. 1 and 2; Harvey 1982, 1985). Surpluses of capital and labor are then absorbed through some mix of long-term growth strategies (usually the debt-financing of long-term investments, which puts off crisis formation into the future) and geographical expansionism (export of money capital, expanding exports of goods and services to other regions, and the like). The problem, of course, is that all other urban economies experience the same dilemmas. To the degree that each strives to rid itself of its own internal contradictions through geographical expansion, the result is economic and geopolitical conflict within the international division of labor (wars over jobs, investment, commodity prices and exchange, money, capital flows, labor migration, and the like). Even in the situation where a particular urban economy has evolved (perhaps through monopolization) toward a point of strongly structured coherence, external influences of this sort also threaten to disrupt it.

We shall return to the geopolitical aspects of this later. For the moment, I simply want to insist upon the power of the tendency toward structured coherence in an urban economy and to insist also that the same processes that push in that direction tend to undermine and disrupt what they produce. So do the internal contradictions of capitalism come home to roost within the economies of urban regions.

IV. PHYSICAL AND SOCIAL INFRASTRUCTURES

The reproduction of both capital and labor power requires a wide range of physical and social infrastructures. These consolidate and reinforce the trend toward structured coherence within an urban labor market.

Some of these infrastructures are embedded in the land as a built environment of roads, bridges, sewers, houses, schools, factories, shopping centers, medical facilities, and so on. They hang together as a spatially specific resource complex of humanly created assets to support both production (fixed capital) and consumption (the consumption fund). They absorb often large quantities of long-term and geographically immobile capital investment and require further capital during their lifetime to compensate for wear and tear and maintenance needs. The aging of this capital stock does not fit any unified schema. It usually has to be renewed piecemeal, under conditions where the relations within the spatial configuration of the resource complex as a whole constrain what can happen to the parts because of the need to preserve the harmony of the whole. The stock of fixed capital and of consumption fund assets does, however, provide a solid form of wealth that can be used to produce and consume more wealth. The urban region acquires another meaning – it can be defined as a particular spatial configuration of a built environment for production, consumption, and exchange.

I have elsewhere examined the general conditions for the production of these kinds of assets and the tensions that exist between them and the dynamics of accumulation (Chap. 1; Harvey 1982, chap. 8). I need state here only those general points necessary to my argument. First, the assets themselves embody or support a dominant technological mix, giving added strength to the idea of structured coherence of production and consumption. Second, privileged access to any unique bundle of assets in the built environment is a potential source of excess profit. Capitalists therefore have a direct interest in the creation and location of such investments and will seek an advantageous location with respect to them (this can lead to strong competitive bidding for sites and locations). Third, the production of the built environment means withdrawing capital from current consumption and production, and that is usually done through debt-financing. Government and financial institutions are usually involved in their production and maintenance. This carries the added advantage that the configuration of the resource complex as a whole can be planned on "rational" lines with an eye to the working harmony of the whole. Fourth, the capital embodied in the built environment is vulnerable to devaluation if the pattern of uses envisaged does not materialize. Put the other way round, this means that employers and consumers are confined to certain kinds of uses and activity patterns for the lifetime of such investments if devaluation is to be avoided. Substitutions of uses are always possible, of course, so that some degree of flexibility exists within these patterns. Finally, the protection of the value of such assets is a vital objective for those who hold the debt on them (primarily financial institutions, governments, and private individuals). Protection in this case means pressure on users to confine themselves to the possibilities that the assets define. In all of these respects, powerful forces are at work to maintain and even institutionalize the structured coherence of the urban-regional economy.

Social infrastructures are harder to pin down. They are not immobile and fixed in space in the same way as built environments (though to the degree they use the latter they also are confined spatially), and they have a variety of orientations (from care of the aged and maintenance of the labor reserve in a state of readiness to enter production, through active policies to enhance labor qualities and ensure discipline and respect for authority, to essential governmental, legal, technical, and scientific services for capitalists). The market area for each kind of service is often vague, and in any case varies in scale from local day care centers to cultural institutions that serve a large region. But they absorb large quantities of capital, and their aggregate effect is to help consolidate the tendency toward structured coherence within the urban region. Furthermore, the social institutions that support life, work, and the circulation of capital are not created overnight and require a certain degree of stability if they are to be effective. The institutions are often national and regional rather than local in scope, but no matter how centralized the degree of financial or political power which lies behind them, some degree of local autonomy is always granted. Social infrastructural provision tends to be hierarchically organized (like labor and commodity markets), with the urban region forming one layer of the structure. Within the urban region, however, institutions and the people who run them tend to coalesce, sometimes tightly and sometimes loosely but rarely without conflict, into a matrix of interlocking and interdependent social resources offering a specific mix of social possibilities. This matrix affects the qualities of labor power in all their aspects (from skills to work attitudes), the condition of the industrial reserve army, and other crucial aspects of labor power supply. But it has equal significance for the reproduction of capital, affecting the production of scientific and technical knowledge and the evolution of managerial and financial know-how and entrepreneurial abilities. Some of the infrastructures are public; some, like education, may be mixed; and still others may be organized outside the framework of the state (religion being a prime example). The reproduction of social infrastructures is therefore open to a curious mix of private and class pressures, social conventions and traditions, and political processes contained within a hierarchically organized state apparatus.

Under such conditions, initial diversities of culture, religion, racial heritage, social attitudes, class consciousness, and so forth can be reproduced and even magnified to form the basis of labor market segmentations. Bourgeois interests may become similarly fragmented through, for example, religious or ethnic domination of economic activity in certain sectors. The effect is to produce further fragmentations in political ideology and allegiance. But taken together, these features give unique coloration to socioeconomic and political processes within each urban region. The social infrastructures are themselves produced by a long history of social interactions and evolution. They are not entirely imposed from above or given from outside. Local industrialists may support scientific and educational institutions in the hope of drawing upon the technologies and managerial skills produced there, while access to such centers (Stanford and MIT, for example) can give competitive advantages. Labor struggles within the urban region can likewise produce strong local commitment to traditions of public education to which the capitalists may also rally in their own class interest. The overall effect of such processes is to emphasize the trend toward structured coherence within the social structure and economy of the urban region.

There are, however, limits to the coherence that can be achieved. The social infrastructures absorb vast quantities of capital and labor power and are limited by the availability of surpluses of both. And to the degree that they become caught up in the circulation of capital (see Chap. 1, fig. 3), so their capacity to enhance local surplus value productivity becomes an important issue. The problem in this case is rendered even more complicated because the benefits of local investments can spread quickly and widely. Interurban "brain drains" of skilled labor and of entrepreneurial and managerial know-how are easily accomplished. New technologies designed in one place can be instantaneously implemented in another. The financing can likewise entail all sorts of redistributions of resources from one urban region to another (through government budgets but also through private transfers, such as alumni donations to colleges). Under such conditions, local exposure to threats of overaccumulation and devaluation becomes hard to estimate. Writing off the value of assets embodied in this resource complex, to say nothing of the human capacities employed there, is, however, a tricky and dangerous affair. Like the built environment, it is hard to change one aspect without affecting the rest. Devaluation entails the modification or sometimes even the destruction of a whole system of community reproduction. Social infrastructures, themselves the product of struggle and history, are hard to transform except through the same kind of creative destruction applied to built environments.

An examination of physical and social infrastructures will help to broaden the conception of what an urban region is all about. It is more than a set of overlapping and interpenetrating commodity and labor markets; more than a set of intersecting labor processes and productive forces; more, even than a simple structured coherence of production and consumption. It is also a living community endowed with certain physical and social assets, themselves the product of a long process of historical development and class struggle. These assets define the wealth of a community, and it is through their proper maintenance, enhancement, deployment, and use that the productivity of labor power stands to be continuously preserved and enhanced at the same time as the reproduction and expansion of capital is assured. The problem, of course, is that this wealth is produced and sustained through the circulation of capital, which is itself crisis-prone. The viability of the physical and social infrastructures is perpetually threatened by the cold winds of overaccumulation, devaluation, and dissolution.

Nevertheless, we now see the urban as a community in which daily processes of living and working occur against a seemingly solid, secure, and relatively permanent background of social and physical infrastructures strongly implanted within the social and physical landscape of capitalism. This implies that a class-bound mode of production and consumption cannot function without some operative geographical conception of community. "Community" is not defined, however, as some autonomous entity but as a set of processes which produce a geographical product. The latter is real and tangible enough. For that reason it leads directly to the question, to what degree is the process set in motion through human agency then dominated by its own product? Put another way, do produced communities act as a barrier to class-bound accumulation? Important political implications then follow.

V. CLASS-ALLIANCE FORMATION IN URBAN REGIONS

The tendency toward structured coherence of the economy gives a material base to class-alliance formation within urban regions. The objective of the class alliance is to preserve or enhance achieved models of production and consumption, dominant technological mixes and patterns of social relations, profit and wage levels, the qualities of labor power and entrepreneurial-managerial skills, social and physical infrastructures, and the cultural qualities of living and working. The class alliance is always unstable (for reasons I shall shortly take up), and its spatial range is in any case fuzzy and usually internally fragmented to some degree (e.g., city versus suburb). Its posture may be defensive or aggressive with respect to other urban regions, but its strength is of particular importance at times of crisis when struggles erupt over the when and where of place-specific devaluation. It can also form alliances with political forces in other urban regions and so build toward regional or national configurations of political power. The class alliance that forms within an urban region is, nevertheless, a powerful shaping force in the landscape of capitalism. The product of capital accumulation and class struggle unfolding in geographical space, it in turn shapes their dynamics in fundamental ways.

Three questions have to be addressed about such class alliances. First, who participates in them and why? Second, how are the diverse interests shaped and articulated politically? And third, what renders such class alliances both unstable and vulnerable?

The short answer to the first question is, everyone but no one in particular. All economic agents occupy a space and have some interest in controlling activities in the spaces around them. If, as we have argued, there is an inevitable tendency toward the production of structured coherence of an urban economy, then it follows that everyone has some interest in finding political means to affect the form that structured coherence takes. But some have a greater interest than others. We know, for example, that capital invested in the built environment cannot be moved without being destroyed. The capital flowing into social infrastructures, though more flexible, is also hard to render mobile without destroying many of its essential qualities. We also know that if the value of this vast capital investment is to be preserved, then production and consumption have to continue at a certain level and of a certain type for a relatively long period. The owners of this capital (or of titles to the public and private debt incurred thereon) have an enormous stake in defending their assets and the models of production and consumption which underpin their value. The ownership of such assets and of the debt can spread widely across social classes, from the working-class homeowner to the large financial institutions that may hold much of the mortgage and municipal debt. All have a vested interest in the continued prosperity of the urban region and have very good reasons to participate in a class alliance to defend their interests. But some factions of capital and labor are more committed by immobile investment than others. Land and property owners (including that faction of the working class that has gained access to homeownership), developers and builders, the local state, and those who hold the mortgage and public debt have much more to gain from forging a local alliance to protect their interests and to ward off the threat of localized devaluation than do transient laborers, itinerant salesmen, and peripatetic multinationals.

But the quantities and qualities of physical and social infrastructures affect the competitive position of the urban region in the international division of labor, the profits of enterprise, the standard of living of labor, and a whole host of other possibilities for the qualities of living and working within the urban region. There are, therefore, broader class interests behind their production and proper utilization than the immediate interests of those who own them. Peripatetic multinationals have a fine appreciation of them. Itinerant tenants are not indifferent to their accommodations either. A broad consensus of interest therefore exists behind the principle (though not necessarily the detailed practices) of enhancing investment in social and physical infrastructures within the urban region (providing, of course, the investment is productive, profitable, and does not unduly favor or burden one class faction rather than another). To the degree that class alliances form within the urban region around the theme of protection and enhancement of immobile physical and social infrastructures, all classes and factions have an interest in participating in the political game.

But interest in class-alliance formation does not end there. Producers who cannot easily move because of fixed capital commitments or who have acquired some degree of monopoly power through privileged access to markets or inputs (including special qualities of the labor supply) can join with an immense array of merchants, professions, services, and state personnel who draw their incomes from the local circulation of revenues to support and preserve the development of the urban economy. Factions of labor that have, through organization and struggle, managed to create islands of relative privilege within a sea of exploitation will rally to such an alliance if its themes are protection of jobs and of living standards already achieved and will see active participation in such an alliance as a means to further enhance their own position. We see the basis here for the rise of some kind of alliance between all classes, in defense of social reproduction (of both accumulation and the reproduction of labor power) within the urban region. The alliance typically engages in community boosterism and strives to create community solidarity behind ideals of social progress and defense of local interests. Such activities, I want to stress, are not aberrations of class struggle but are a necessary and particular manifestation of the way class relations and accumulation unfold in space.

Such alliances are, however, inherently unstable. Both internal divisions and external pressures make it hard to hold them together in the face of a social dynamic restlessly powered by the pursuit of profit, the accumulation of capital, and the multifaceted lines of class cleavage and struggle embodied therein. Divisions become immediately apparent when it comes to mapping the future. Different interests pull in different directions, each usually claiming that the public interest lies wherever it itself is headed. Factional divisions within the bourgeoisie (between financial, commercial, producer, real estate, and landed property interests, or between local neighborhood producers and multinational organizations) match factional divisions within the working class (between men and women, between skilled blue collar and white collar and the unskilled, between the employed and the unemployed, between the varied segmentations) to make it hard to talk of any coherent class interest in class-alliance formation. Worse still, individuals occupy multiple roles and can be torn in many different directions – workers may also be homeowners, consumers, parents, and investors and may seek to participate in the class alliance in quite inconsistent roles. And there is no way a class alliance can act that does not unduly favor or burden one faction or class rather than another. Decisions on public investment, to take the clearest case, have uneven class impacts and benefits at the same time as they alter the spatial configuration of assets and their relative accessibility. Work-force segmentations that have become manifest as spatial segregations stand to be undermined or reinforced, depending upon the nature of such decisions. Competition (between workers, between producers of goods and services, retailers, etc.) and the struggle to procure monopoly powers do not disappear with class-alliance formation. They are, rather, perpetually disruptive forces that the class alliance has to contain. Much of the art of urban and regional politics, as we shall shortly see, consists in finding ways to trade off costs and benefits between groups and interests while containing competition and monopoly powers so as to maintain majority support for a ruling alliance.

The external pressures on the alliance's stability are of two sorts. First, all economic agents internalize a choice between staying in place and striving for local improvement or moving elsewhere to where profits, wage rates, working conditions, life-styles, environmental qualities, hopes for the future, and so on appear better. That tension, common to all, is not evenly balanced for all. Different factions and classes have different capacities for geographical mobility depending upon the privileges they command, the assets they own, and the intangible restraints that tend to keep them place-bound. A single male with a sack of gold (or, what amounts to the same thing, letters of credit) has more options as a rule than does the owner of a local steel mill or a married woman with extensive family and kinship ties and cares. Some are more solid partners in a class alliance than others simply because they have fewer options to move elsewhere. But appearances can sometimes be deceptive. Bankers and financiers control the most geographically mobile asset of all – money – but are also often heavily committed to an urban region through their holdings of local debt (this was the dilemma of many of New York's international banks during that city's fiscal crisis of 1974–75). Multinational firms appear able to relocate production rapidly, yet sometimes depend upon such a particular mix of fixed capital, local labor qualities, and infrastructures that movement comes hard. A worker with extensive kinship obligations may use them to command geographical mobility rather than to remain locked in place. These examples illustrate that decisions to support or abandon, build or undermine a class alliance must come out of the resolution of complex and conflicting tensions. The same group interest may even actively undermine on the one hand what they actively strive to support on the other. Financial institutions, for example, may undermine the quality of their own debt and the power of their local class alliance by financing suburbanization or the export of money capital to wherever the rate of return is highest. Workers, by pushing for high wages, can stimulate the loss of jobs. Such examples of unintended and often contradictory consequences are legion in this context.

Second, disruptive forces can be brought to bear on the class alliance from outside. The in-migration of labor power of lower cost and different qualities, the takeover of local production and retailing by outside capital, the import of commodities once locally produced, inflows of money capital, and redistributions of revenues alter power balances between the participants in an alliance. Indeed, the import may be organized by one faction with just such an aim: employers encourage in-migration of cheaper labor power; merchants, the import of cheaper commodities compared to those produced locally; and so on. The ability to mobilize external relations and possibilities becomes an important bargaining strength in negotiations within the class alliance. The class or faction that can most easily summon up external assistance (the labor unions that can bring in strike support funds, the capitalists who can mobilize outside support to quell unrest) have an advantage over groups who lack such power. Conversely, groups that can threaten to move elsewhere if they do not get local satisfaction of their demands are in a more powerful position relative to those that cannot.

The same forces that counteract the tendency toward a structured coherence of an urban economy also render class alliances unstable and insecure. But there is a further dimension to all of this, which we must now subject to rather close scrutiny. This concerns the political means available to define, articulate, and act upon class-alliance aims and the political art of forging a ruling-class alliance out of the conflictual and contradictory impulses that lie behind the tendency toward local class-alliance formation.

VI. URBAN POLITICS AND THE SEARCH FOR A RULING-CLASS ALLIANCE

The confusions and instabilities of class-alliance formation create a political space in which a relatively autonomous urban politics can arise. The confusions of roles, orientations, and interests of individuals, groups, factions, and classes, taken together with the disruptions of capital accumulation (growth, technological change, class conflict, and crises of overaccumulation), keep social relations in a perpetual state of flux and often plunge them into the ambiguous tensions of social transformation. The art of politics here comes into its own. The politician of genius and craft can forge a relatively permanent and powerful coalition of interests so as to unify and articulate a sense of place-bound community. Indeed, so open is the situation that a whole class of politicians can arise given over entirely to its exploitation. "Nowhere do 'politicians' form a more separate, powerful section of the nation than in North America," wrote Engels, pushing the "process of the state power making itself independent in relation to society" to extremes (Marx and Lenin 1968, 20). It is into this breach that a whole class of "urban managers" can also insert themselves as a seemingly independent source of social power (Saunders 1981, 118–36). Both politicians and urban managers (and there often seems little point in distinguishing between them from this standpoint) play the game of coalition politics in such a way as to build a ruling class that sees itself as the symbol of community and appropriates the necessary means (traditional and symbolic as well as legal, financial, and technical) to legitimize its authority and power. It usually speaks "in the public interest" and finds ways to command sufficient authority or mass support (by way of concessions, cooptation, horse-trading, and repression) so as to still the opposition that is bound to arise to its activities.

The local government is, of course, a central political means around which a ruling coalition tends to forge its identity and modes of action within an urban region. But I want to resist the idea that it is the only or even the most important means. The political processes at work in civil society are much broader and deeper than the local government's particular compass. Indeed, there are many facets that make it ill-suited to the task of coalition building. Its boundaries do not necessarily coincide with the fluid zones of urban labor and commodity markets or infrastructural formation; and their adjustment through annexation, local government reorganization, and metropolitan-wide cooperation is cumbersome, though often of great long-run significance. Local jurisdictions frequently divide rather than unify the urban region, thus emphasizing the segmentations (such as that between city and suburb) rather than the tendency toward structured coherence and class-alliance formation. Other means then have to be found within the higher tiers of government or within civil society (informal groupings of business and financial interests, for example) to forge a ruling-class alliance. On the one hand, Robert Moses reshaped New York without any popular mandate by using state and federal powers, backed by a network of powerful financial, business, construction (including unions), and real estate interests, to dominate an otherwise segmented local government apparatus. Mayor Schaefer uses Baltimore's City Hall, on the other hand, as a base to reach out into civil society and build a coalition of public and private interests capable of dominating the whole urban region. It is, we conclude, the interpenetration of class, group, and individual relations within and between the state and civil society which provides the matrix of possibilities for building a ruling coalition.

To the degree that all economic agents have some interest in joining a ruling coalition, the composition of the latter is open rather than predetermined. Its exact composition is a matter of negotiation out of which many different configurations can arise. Alliances can shift from issue to issue (capital and labor may agree upon the need for new jobs but disagree about the need to regulate working conditions), and different working coalitions define varied and sometimes quite contrary objectives. Some coalitions may be pro-growth and others anti-growth, and elements of capital and labor can be found on both sides of that divide. And the politics can point in many different directions. On the one hand, there are the urban-based revolutionary movements such as that of the Paris Commune, and the strong traditions of "municipal socialism" – Milwaukee in the 1900s; Vienna in the 1920s;

Bologna, the Greater London Council, and Santa Monica today – which sustain themselves through electoral command of the local government apparatus. On the other hand, there are the seemingly all-powerful pro-growth coalitions that emerged in many American cities after 1945, in some cases using and in other cases by-passing (like Robert Moses) local government (Molotch 1976; Mollenkopf 1983). In between lie all kinds of hybrids from the hotly contested politics described in Katznelson's (1982) *City Trenches* to the stable but authoritarian machine politics of New York's Boss Tweed, Chicago's Mayor Daley, and Baltimore's Mayor Schaefer. Each kind of politics depends upon the forging of a particular coalition of interests; and each is, in its own way, unique. Furthermore, each coalition has different means and resources open to it which limit what it can or cannot do (political control over the state apparatus, the local budget, and land-use regulation, for example, gives very different powers from control over the strings of investment finance). Herein lies the sort of tension that sparks conflict and that can bring a ruling coalition down. When Mayor Kucinich tried to take the local government apparatus of Cleveland against the banking community, he eventually lost, to be replaced by a new ruling-class alliance in which financier and City Hall cooperated.

The impact of the ruling coalition upon the pace of local growth, innovation, social transformation and reproduction can be far-reaching and profound in its implications. Not only can it exercise direct control over the formation of physical and social infrastructures (and through them influence the basic economic and social attributes of the urban region), but it can also go out of its way to attract or repulse jobs (of this or that sort), people (of this or that class or sort), and business, commercial, financial, real estate, cultural, and political activities. It can strive to create an appropriate "business climate," fashion new kinds of living environments, encourage new kinds of life-style, facilitate and attract new kinds of development. It can be innovative or defensive, passive or aggressive in its pursuit of social objectives and economic goals. Even new patterns of social relations can be affected – segmentations of one sort may be diminished (for example, between the races and sexes), while discrimination of another sort can be highlighted (for example, a privileged and politically conscious faction of labor may be detached from the rest of the working class by acquisition of special privileges within a ruling-class alliance). From all of these standpoints, the political-economic evolution of an urban region appears relatively autonomous and certainly unique and particular to every instance.

The multidimensional ferment and unique qualities of such political processes within the urban region, the forging of unique ruling coalitions out of all kinds of individual, group, and class fragments, the unique directions taken, and the powerful mobilization of the spirit of a place-bound geographical community appear at first sight as quite incompatible with the basic presuppositions of a capitalist mode of production and consumption. But there is a deep sense in which such features are not only compatible with but integral to capitalism's processes and contradictions as these necessarily operate in geographical space. It is important to see how and why.

Any ruling-class alliance has to accommodate to the basic logic of capital circulation and accumulation if it is to remain within the capitalist system and successfully reproduce the conditions of its own existence. A successful ruling-class alliance has to be, in spirit as well as in practice, a procapitalist class alliance. The trouble, of course, is that there are many ways to be procapitalist, while the inner contradictions of capitalism render any attempt to be consistent moot. Being procapitalist certainly does not mean selling out to a local capitalist class, since such groups do not necessarily act in their own class interest any more than individuals do (particularly when these groups exert some degree of monopoly control). And when we introduce the uncertainties of spatial competition under geographical conditions of changing space relations, it becomes evident that no single line of argument or action can define what it means exactly to be procapitalist. Even if capitalists mounted a powerful conspiracy (and from time to time they do), the odds are that it would not work.

This is the kind of situation which Marxists find so discouraging and their critics delight in. It underlies all the debates over the virtues of class analysis versus urban managerialism, over the "relative autonomy" of local political processes, over "place as historically contingent process," as Pred (1984) termed it, versus a general theory of uneven capitalist development in geographical space. How, then, can we cut the Gordian knots in these tangled debates?

VII. THE URBAN REGION AS A GEOPOLITICAL UNIT IN THE UNEVEN GEOGRAPHICAL DEVELOPMENT OF CAPITALISM

Daily life is reproduced under capitalism through the circulation of capital. That circulation process has a certain contradictory logic, entails class relations and struggle, promotes perpetual revolutions in productive forces and modes of consumption, and requires a mass of supporting organization and infrastructure to reproduce itself. We know a great deal about that contradictory logic (cf. Harvey 1982). The problem is to show how the phenomena of the urban process are contained within it. I say "contained within" rather than "reduced to" precisely because I regard any account of the circulation of capital as incomplete that does not include, among other things, its geographical specification. Though I will concede, therefore, that there are aspects of urban life and culture which seem to remain outside the immediate grasp of the contradictory logic of accumulation, there is nothing of significance that lies outside its context, not embroiled in its implications. The task of the urban theorist, therefore, is to show where the integrations lie and how the inner relations work.

Capital accumulation, when considered as a geographical process from the very start, tends to produce distinctive urban regions within which a certain structured coherence is achieved and around which certain class alliances tend to form. Conceding the instability of that process (including the instability of the geographical space and its definition) opens a space for seemingly autonomous political processes and for seemingly unique ruling coalitions to form, taking each urban region down a distinctive political-economic path of development. I now have to show how and why the autonomy and the uniqueness are not only compatible with but also vital to the logic of accumulation in geographical space. Once the question is posed that way, it proves not too hard to at least sketch in an answer. I shall do so through consideration of four basic points.

Consider, first, the idea of "the city as a growth machine." I use Molotch's (1976) telling image in part because it reflects the capitalistic imperative of "accumulation for accumulation's sake, production for production's sake." But deeper consideration of it takes us past the mere convenience of analogue and into more fruitful theoretical territory. Accumulation entails the conversion of surplus capital through combination with surplus means of production and surplus labor power into new commodity production. That activity is inherently speculative. But accumulation also requires the prior production of the necessary preconditions of production, the social and physical infrastructures being of the greatest significance in this regard. The production of these preconditions by capital entails a double and compound speculation. Prior speculative investments have to match the requirements of further speculative growth. And these prior investments are at least in part embedded in the land as immobile and fixed capital of long duration. For the individual capitalist, of course, the most convenient condition is that in which they can either freely appropriate prior conditions as they find them (for example, assets generated under some prior mode of production) or make minimal investments on their own account (a rail connection, some worker housing, a company store). But that is insufficient for sustained accumulation. The politics then have to precede the economy.

It is at just such a point that a ruling coalition and the autonomy of its politics come into their own. A ruling coalition in effect speculates on the production of the preconditions for accumulation; it collectivizes risks through finance capital and the state. This is precisely what the "growth machine" is all about. Yet it is, as Molotch insists, a *capitalistic* growth machine in which certain dominant interests – of banking and finance capital, of property capital and construction interests (including laborers and their unions), of developers and ambitious agents of the state apparatus – typically call the tune. They seek profit from the production of preconditions. The realization of that profit depends on the profitability of the accumulation that such preconditions help promote. The growth coalition uses its political and economic power to push the urban region into an upward spiral of perpetual and sustained accumulation. Such a process has its inner tensions and conflicts and cannot, given the contradictory logic of capital accumulation, be permanently sustained. We shall return to its instabilities shortly.

Consider, second, invention and innovation. Intercapitalist competition and class struggle force periodic revolutions in productive forces. Such conditions vary from one urban region to another. But the search for excess profit also spurs innovation for innovation's sake as well as attempts to counter that thrust through monopoly controls and locational shifts. Jane Jacobs (1969, 1984) has long argued, for example, that the fundamental role of cities is to produce "new work" and that some cities are better at it than others. Those with chaotic industrial and entrepreneurial structures allow of the unexpected collision of new ideas, techniques, and possibilities out of which new products and methods can spring. Those in which monopoly power is deeply entrenched are less open and more prone to stagnation. That thesis, partly plausible for the nineteenth century, is less so today. If innovation has become a business (as Marx long ago argued it must), then the creation of preconditions of that business becomes more and more important. These preconditions can be more easily sustained within the large multinational corporation and the state than by the small firm. And while spatial agglomeration of such preconditions (the "high-tech" innovation centers around Boston, Palo Alto, North Carolina, Long Island, etc.) may be relevant, the spatial transfer of technology, albeit often under monopoly control, is now so rapid as to render its specifically urban qualities moot. Yet there is a broader version of Jacobs's thesis that makes more, though still only partial, sense.

Innovation, after all, entails more than invention. It calls for venture capital and specific labor skills in its development, access to distribution systems for marketing, and openness on the part of recipients which may entail the redesign of consumer markets and the transformation of taste and fashion. It affects the hegemonic technological mix within the structured coherence toward which every urban economy tends. Innovation, in short, is not and can never be confined to the sphere of production. It necessarily spills over into consumption, household reproduction, social services (e.g., education, health care), administration, cultural activities, and political processes. There is also a strong demand for it in the military and in other branches of government concerned with surveillance and control. Innovation in all these spheres is as important to the dynamics of capitalism as are direct changes in the labor process. This social and political innovation has to be "rational," however, in relation to accumulation. How is such a result achieved? From this standpoint we can view the urban region as a social and political innovation center within which the search for some appropriate mix of life-styles, social provision, cultural forms, and politics and administration parallels the perpetual thrust toward technological and organizational dynamism in production. The autonomy of the urban region's ruling-class alliance and of its politics is vital to this kind of social and political dynamism. Furthermore, the liberty of individuals and groups to intervene in that politics is as fundamental as is the liberty of entrepreneurs to pursue technological changes and product innovation. The social ferment and conflict of urban social movements born out of class struggle, possessive individualism, community rivalries, and segmentations and segregations based on labor qualities and life-style preferences can be mobilized into creative processes of sociopolitical innovation. The successful urban region is one that evolves the right mix of life-styles and cultural, social, and political forms to fit with the dynamics of capital accumulation.

But how do we know when the right mix is achieved? This brings me to a third point. We can view the urban region as a kind of competitive collective unit within the global dynamics of capitalism. Like individual entrepreneurs, each urban region has the autonomy to pursue whatever course it will, but in the end each is disciplined by the external coercive laws of competition. Its industry has to compete within an international division of labor, and its competitive strength depends upon the qualities of labor power; the efficiency and depth of social and physical infrastructures; the "rationality" of life-styles, cultures, and political processes; the state of class struggle and social tension; and geographical position and natural resource endowments. Urban regions that make wrong choices lose out to their competition in much the same way that erring entrepreneurs do. Urban regions wracked by class struggle or ruled by class alliances that take paths antagonistic to accumulation (toward no-growth economies or municipal socialism) at some point have to face the realities of competition for jobs, trade, money, investments, services, and so forth. Urban regions can be left behind, stagnate, decay, or drift into bankruptcy, while others surge ahead. But this is not to say that all kinds of successful specializations, particular mixes of urban economy and divisions of labor, ruling-class alliances and divergent political forms, cannot coexist. The uniqueness of each urban region is not eliminated by capitalism any more than the individual firm loses its unique qualities. Some urban regions specialize in the production of surplus value, while others seem to specialize in consuming it. Some appear at a certain historical moment as leading centers of cultural and political innovation, only to fade under the heavy hand of some dominant ruling class that so stifles dissent that innovation lags. Other class alliances use strong coercive powers to force a recalcitrant population into the forefront of accumulation through the disciplining of labor movements and the reduction of wage rates and worker resistance. All kinds of combinations are possible. But the uniqueness has to be seen as historically and geographically contingent. The combinations, arrived at through voluntaristic and autonomous struggles, are in the end contingent upon processes of capital accumulation and the circulation of associated revenues in space and time.

But now I shall modify that conception somewhat through consideration of a fourth point. The political power of a ruling-class alliance is not confined to an urban region: it is projected geopolitically onto other spaces. We now have to see the urban region as a geopolitical entity within the uneven geographical development of capitalism. How that geopolitical power is projected and used has important consequences, not only for the fate of the individual urban region but for the fate of capitalism. Let us see how that can be so.

The power that a ruling-class alliance projects depends in part upon the internal resources it can mobilize. Financial and economic leverage is crucial. That in part depends upon the urban region's competitive position. But competition is not always between equals: urban regions with enormous and complex economies cast a long and often dominant shadow over the spaces that surround them. Economic power is deployed within a hierarchical structure of urban regions. Those urban regions, like New York and London, which command power within the realms of credit and finance (the central nervous system of capitalism) can use that power across the whole capitalist world. The hierarchy of size is reemphasized by hierarchies of function. The power of innovation, in social and political affairs as well as in the production of goods and services, also confers a particular influence. But political leverage within a hierarchically organized state apparatus is also important in its own right. Power, in this case, depends largely upon the coherence and legitimacy of the local ruling coalition in relation to national politics.

These different sources of local power are not always consistent with each other. Divisions and fragmentations abound and frequently check the geopolitical influence of the urban region as a coherent entity. The conflict that attaches to social and political innovation, for example, may fragment and divide a local alliance, leaving it open to external influence and manipulation. Capitalists may appeal to capitalists in other regions or to state authority to put down labor unrest, while workers may likewise build coalitions across urban regions and seek command of central state power for their own advantage. Whoever is excluded from a local ruling coalition will likely seek outside help. Different coalitions also command different resources. A local socialist movement may have a great deal of local legitimacy and popular support but lack command over the levers of finance and trade. The exercise of geopolitical power may under such circumstances split into two or more factions: the "city" of London has a far different kind and range of influence than the Greater London Council. The fragmentations can just as easily be geographical as social. Political jurisdictions can be defined, for example, which emphasize the split between city and suburb.

We here encounter the fascinating issue of the scope and extent of local political authority in relation to economic interests within the urban region. The trend in the nineteenth century was to try to extend and enhance local political authority so as to make the major cities geopolitical entities that reflected the main lines of economic influence and power. But after universal suffrage and the growth of labor movements, the trend has been in the opposite direction, toward political fragmentation of urban regions and cleavage between economic and political scales of operation. The problem for practical urban political economy is, therefore, to establish the urban region as a coherent geopolitical presence in the face of the fragmentations. To this end, a ruling-class alliance will seek to mobilize sentiments of community boosterism and solidarity, to coopt and create loyalties to place, to invent and appropriate local tradition. It will use local newspapers, radio, and television to reinforce the sense of place (this process is particularly powerful in the United States) in a world of universal exchange. It will strive to build a political machine, not necessarily confined to conventional political channels, to wage geopolitical struggle in a world of uneven geographical development.

Geopolitical strategies become part of the arsenal of weapons employed by a ruling coalition. At the very minimum, the coalition will struggle politically within the nation state over the allocation of public investments, tax incidence and revenues, political representation, and the like. It will seek ways to enhance its geographical position through political concessions and public investments (particularly in transport and communications, in which regard the ruling coalition tends to act as a glorified collective land speculator). Building a powerful political machine able to deliver the vote and other kinds of political support often pays off in political and economic favors. But economic, cultural, and innovative powers can also be used as instruments of domination over other urban regions. Surpluses of capital and labor generated within the urban region can be put to use elsewhere under controlled conditions; branch plants can be set up in different urban regions, exclusive trading outlets cultivated, and powerfully dominant financial links forged. Tentacles of economic power can reach outward to dominate other urban regions, while invading tentacles from elsewhere are vigorously combated by political maneuvers and monopolistic controls. Competition between urban regions is thus transformed into a raw geopolitical power struggle between them.

The ruling coalition has, after all, much to defend in the way of sunken investments, standards of living, conditions of work, life-styles and culture, social organization, and modes of politics and governance. It also has much to gain from rapid and often conflictual innovation in all these areas. When times are good, it can look to improve its competitive position and bring the fruits of capitalist "progress" into the community. It can look to mobilize internal and external forces into an upward spiral of local development. In times of crisis it has just as much to gain by warding off the devaluation and destruction of productive capacity, labor power, local markets, and social and physical infrastructures. It can look to mobilize competitive and geopolitical power to export the threat of local overaccumulation and to bar the import of such problems onto its own terrain.

In all of these respects the ruling coalition has to act like a kind of collective entrepreneur. Its role is double-edged. Competition between different urban regions and the coalitions that represent them helps co-ordinate the political and social landscape with capitalism's exacting requirements. It helps discipline geographical variations in accumulation and class struggle within the bounds of capitalism's dynamic at the same time as it opens fresh spaces and possibilities within which that dynamic can flourish. The different coalitions become key agents in the uneven geographical development of capitalism. Insofar as that uneven geographical development, as I have argued elsewhere (Harvey 1982, chaps. 12 and 13), is a stabilizing outlet for capitalism's contradictions, so the agency that helps promote it becomes indispensable. Capitalism's pursuit of a spatial fix for its own inner contradictions is actively mediated through the actions of a ruling coalition attentive to the fate of accumulation and class struggle within each urban region. But like all entrepreneurs, the coalition is caught between the fires of open and escalating competition with others and the stagnant swamp of monopoly controls fashioned out of the geopolitics of domination. The latter entails the crystallization of the geographical landscape of capitalism into stable but in the long run stagnant configurations of hierarchical domination within a system of urban regions. The problem, of course, is that capitalism cannot so easily be contained; stagnation is not its forte but merely compounds and exacerbates internal contradictions. The coherence of a ruling coalition is put at risk under such conditions, at the very moment when external opposition to its domination necessarily hardens. The collective entrepreneur, we must recall, is fashioned out of an uneasy and unstable coalition of individuals, factions, and classes, each of which internalizes a tension between seeking advantage by breaking from or even undermining the coalition and remaining solidary and so seeking to secure gains already made. The fragments are, in any case, always caught between postures of conflict and accommodation. To the degree that a ruling coalition does not deliver on its promises – as must in the long run be the case under monopoly control – so the forces making for its overthrow, always latent, move explicitly into open and sometimes even violent revolt. The weakening of internal coherence gives abundant opportunity to reshape external links and alliances as well as to shape new internal combinations of forces. New urban regions arise as power centers within the international division of labor; innovation and competitive growth resume at the same time as the forces of competitive restructuring through creative destruction are put to work. The geopolitical landscape of capitalism, like that of production, "lurches between the stabilizing stagnations of monopoly control and the disruptive and often destructive dynamism of competitive growth."

That process is stressful in the extreme. And it provokes its own particular forms of resistance, sometimes spilling over into revolts against the very logic of capitalism itself. Urban regions are constructed as communities replete with traditions of labor market behavior, capitalist forms of action, class alliances, and distinctive styles of politics. These constructions, built out of the class relations of capital accumulation, are not so easily overthrown. The ruling-class alliances, the urban political processes, and the geopolitical rivalries that capitalism produces therefore appear, at a certain point, as major barriers to further capitalist development. The revolutionary power of capitalism has to destroy and reshape the sociopolitical forms it has created in geographic space. The seeming autonomy and perpetual ferment of urban political processes lies at the very heart of that contradiction. But then, so too does the potentiality of urban politics to shape the cutting edge of revolution. The urban region either submits to the forces that created it or becomes the hearth of a revolutionary movement.

VIII. CONCLUSION

I have long argued that capitalism builds a physical and social landscape in its own image, appropriate to its own condition at a particular moment in time, only to have to revolutionize that landscape, usually in the course of crises of creative destruction at a subsequent point in time (cf. Chaps. 1 and 2). We now encounter a particular version of that thesis, worked out on a particular geographical scale, that of urban regions, and with a much stronger political content. I do not, however, want for one moment to give the impression that this is the only scale on which such a geopolitical representation can be constructed. I have elsewhere sought to represent it from the standpoint of larger-scale regions and nation states (Harvey 1982, 1985). But observers of the urban process, of no matter what political or methodological persuasion, at least agree on this: that social, economic, and political processes have a particular meaning at the urban level of analysis and that such a scale of generalization has real implications for the way in which individuals and other economic agents relate daily actions to global processes. The urban realm is, as it were, a "concrete abstraction" that reflects how individuals act and struggle to construct and control their lives at the same time as it assembles within its frame real powers of domination over them. Urban politics is a realm of action which individuals can easily understand and to which individuals can immediately relate. The sorts of processes we have studied in this chapter provide a real context of conditions to which individuals, groups, and classes must accommodate and respond. The processes appear as abstract forces, to be sure, but they are not the kind of forces that we can ever afford to abstract from. To argue, for example, that class struggle can unfold independently of geopolitical representations and confrontations is a totally unwarranted abstraction. It is, unfortunately, an abstraction to which Marxists have all too frequently been prone. From this standpoint, the critics and defectors from the Marxist tradition of urban analysis are correct in the complaints they voice. Where they in turn err is in seeking to deny that the urban community and its distinctive politics are produced under capitalist relations of production and consumption as these operate in and on geographical space.

The fundamental Marxist conception, as I see it, is of individuals and social groups, including classes, perpetually struggling to control and enhance the historical and geographical conditions of their own existence. How they struggle – individually or collectively, through coalitions or confrontations – has important implications. But we also know that the historical and geographical conditions under which they struggle are given, not chosen. And this is true no matter whether the conditions are given by nature or socially created. The relevant conditions can be specified many ways, however, and some ways appear more relevant than others. Where we put the emphasis matters. In a capitalist society, we know that social life is reproduced through the circulation of capital, which implies class relations and struggle, accumulation and innovation, and periodic crises. But we also have to say something more concrete about the historical and geographical conditions of that process. Marxists have paid close attention to the history – the creation of wage labor, the rise of money forms and commodity production, the formation of necessary social and institutional supports, the emergence of certain kinds of political theory and authority, the constructions of scientific and technical understandings, and the like. But they have paid little attention to the geography. Putting the geography back in immediately triggers concern for the urbanization of capital as one of the key conditions under which struggles occur. It also focuses our attention on how capitalism creates spatial organization as one of the preconditions of its own perpetuation. Far from disrupting the Marxian vision, the injection of real geographical concerns enriches it beyond all measure. By that path we might hope to liberate ourselves from the chains of a spaceless Marxist orthodoxy as well as from the futility of bourgeois retreat into partial representations and naïve empiricism. The stakes of historical and geographical struggle are far too great to allow the luxury of such retreats. The historical geography of capitalism has to be the central object of theoretical enquiry in the same way that it is the nexus of political action.

7

On Planning the Ideology of Planning

It is a truism to say that we all plan. But planning as a profession has a much more restricted domain. Fight as they might for some other rationale for their existence, professional planners find themselves confined, for the most part, to the task of defining and attempting to achieve a "successful" ordering of the built environment. In the ultimate instance the planner is concerned with the "proper" location, the appropriate mix of activities in space of all the diverse elements that make up the totality of physical structures – the houses, roads, factories, offices, water and sewage disposal facilities, hospitals, schools, and the like – that constitute the built environment. From time to time the spatial ordering of the built environment is treated as an end sufficient unto itself, and some form of environmental determinism takes hold. At other times this ordering is seen as a reflection rather than a determinant of social relations, and planning is seen as a process rather than as a plan – and so the planner heaves himself away from the drawing board to attend meetings with bankers, community groups, land developers, and the like, in the hope that a timely intervention here or a preventive measure there may achieve a "better" overall result. But "better" assumes some purpose which is easy enough to specify in general but more difficult to particularize about. As a physical resource complex created out of human labor and ingenuity, the built environment must primarily function to be useful for production, circulation, exchange, and consumption. It is the job of the planner to intervene in the production of this complex composite commodity and to ensure its proper management and maintenance. But this immediately poses the question, useful or better for what and to whom?

I. PLANNING AND THE REPRODUCTION OF THE SOCIAL ORDER

It would be easy to jump from these initial questions straight away into some pluralistic model of society in which the planner acts either as an arbitrator or as a corrective weight in the conflicts amongst a diversity of interest groups, each of which strives to get a piece of the pie. Such a jump leaves out a crucial step. Society works, after all, on the basic principle that the most important activity is that which contributes to its own reproduction. We do not have to enquire far to find out what this activity entails. Consider, for example, the various conceptions of the city as "workshops of industrial civilization," as "nerve centers for the economic, social, cultural and political life" of society, as centers for innovation, exchange, and communication, and as living environments for people.[1] All of these – and more – are common enough conceptions. And if we accept one or all of them, then the role of the planner can simply be defined as ensuring that the built environment comprises those necessary physical infrastructures which serve the processes we have in mind. If the "workshop" dissolves into a chaos of disorganization, if the "nerve center" loses its coherence, if innovation is stymied, if communication and exchange processes become garbled, if the living conditions become intolerable, then the reproduction of the social order is in doubt.

We can push this argument farther. We live, after all, in a society that, for want of a better phrase, is founded on capitalist principles of private property and market exchange, a society that presupposes certain basic social relationships with respect to production, distribution, and consumption which themselves must be reproduced if the social order is to survive. And so we arrive at what may appear a rather cosmic question: what is the role of the urban-regional planner in the context of these overall processes of social reproduction? Critical analysis should reveal the answers. Yet it is a measure of the failings of contemporary social science (from which the planning literature draws much of its inspiration) that we have to approach answers with circumspection as well as tact should we dare to depart from the traditional canons as to what may or may not be said. For this reason I shall begin with a brief digression in order to open up new vistas for discussion.

When we consider the economic system, most of us feel at home with analyses based on the categories of land, labor, and capital as "factors" of production. We recognize that social reproduction depends upon the perpetual combination of these elements and that growth requires the recombination of these factors into new configurations that are in some sense more productive. These categories, we often admit, are rather too abstract, and from time to time we break them down to take account of the fact that neither land nor labor is homogeneous and that capital can take productive (physical) or liquid (money) form. Nevertheless, we seem prepared to accept a high level of abstraction, without too much questioning as to the validity or

[1] These various conceptions of the city can be found in, for example, L. Mumford (1961); J. Jacobs (1969); L. Wirth (1964); National Resources Committee (of the United States) (1937); and R. Meier (1962).

efficacy of the concepts employed. Yet most of us blanch when faced with a sociological description of society which appeals to the concept of class relations between landlords, laborers, and capitalists. If we write in such terms, we will likely be dismissed as too simplistic or as engaging in levels of abstraction which make no sense. At worst, such concepts will be regarded as offensive and ideological compared to the supposedly nonideological concepts of land, labor, and capital. Why and on what grounds, philosophical, practical, or otherwise, was it decided that one form of abstraction made sense and was appropriate while the other was out of order? Does it not make reasonable sense to connect our sociological thinking with our economics, albeit in a rather simplistic and primitive way? Does it make sense even, to tell the inner-city tenant that the rent paid to the landlord is not really a payment to that man who drives a big car and lives in the suburbs but a payment to a scarce factor of production? The "scientization" of social science seems to have been accomplished by masking real social relationships – by representing the social relations between people and groups of people as relations between things. The reification implied by this tactic is plain enough to see, and the dangers of reification are well known. Yet we seem to be at ease with the reifications and to accept them uncritically, even though the possibility exists that in so doing we destroy our capacity to understand, manage, control, and alter the social order in ways favorable to our individual or collective purposes. In this chapter, therefore, I seek to place the planner in the context of a sociological description of society which sees class relations as fundamental.

II. CLASS RELATIONS AND THE BUILT ENVIRONMENT

In any society the actual class relations that exist are bound to be complex and fluid. This is particularly true in a society such as ours. The class categories we use are not regarded as immutable. And in the same way that we can disaggregate land, labor, and capital as factors of production, so we can produce a finer mesh of categories to describe the class structure. We know that land and property ownership comprises residual feudal institutions (the church, for example), large property companies, part-time landlords, home-owners, and so on. We know also that the interests of rentier "money capitalists" may diverge substantially from the interests of producers in industry and agriculture and that the laboring class is not homogeneous because of the stratifications and differentials generated according to the hierarchical division of labor and various wage rates. But in a short chapter of this sort I must perforce stick to the simplest categories that help us to understand the planner's role within the social structure. So let us proceed with the simplest conception we can devise and consider, in turn, how each class or fraction of a class relates to the built environment, which is the primary concern of the planner.

1. The class of laborers is made up of all of those individuals who sell a commodity – labor power – on the market in return for a wage or salary. The consumption requirements of labor – which are in practice highly differentiated – will in part be met by work within the household and in part be procured by exchanges of wages earned against commodities produced. The commodity requirements of labor depend upon the balance between domestic economy products and market purchases as well as upon the environmental, historical, and cultural considerations that fix the standard of living of labor. Labor looks to the built environment as a means of consumption and as a means for its own reproduction and, perhaps, expansion. Labor is sensitive to both the cost and the spatial disposition (access) of the various items in the built environment – housing, educational and recreational facilities, services of all kinds, and so on – which facilitate survival and reproduction at a given standard of living.

2. We can define capitalists as all those who engage in entrepreneurial functions of any kind with the intent of obtaining a profit. As a class, capitalists are primarily concerned with accumulation, and their activities form, in our kind of society, the primary engine for economic development and growth. Capital "in general" – which I use as a handy term for the capitalist class as a whole – looks to the built environment for two reasons. Firstly, the built environment functions as a set of use values for enhancing the production and accumulation of capital. The physical infrastructures form a kind of fixed capital – much of which is collectively provided and used – which can be used as a means of production, of exchange, or of circulation. Secondly, the production of the built environment forms a substantial market for commodities (such as structural steel) and services (such as legal and administrative services) and therefore contributes to the total effective demand for the products that capitalists themselves produce. On occasion the built environment can become a kind of "dumping ground" for surplus money capital or idle productive capacity (sometimes by design, as in the public works programs of the 1930s), with the result that there are periodic bouts of overproduction and subsequent devaluation of the assets embedded in the built environment itself. The "wavelike" pattern of investment in the built environment is a very noticeable feature in the economic history of capitalist societies.[2]

[2] Economic cycles and in particular those associated with investment in the various components of the built environment are discussed in B. Thomas (1973); M. Abramovitz (1964); S. Kuznets (1961); and E. Mandel (1975). See also Chap. 1.

3. A particular faction of capital seeks a rate of return on its capital by constructing new elements in the built environment. This faction – the construction interest – engages in a particular kind of commodity production under rather peculiar conditions. Much of what happens in the way of construction activity has to be understood in terms of the technical, economic, and political organization of the construction interest.

4. We can define landlords as those who, by virtue of their ownership of land and property, can extract a rent (actual or imputed) for the use of the resources they control. In societies dominated by feudal residuals, the landlord interest may be quite distinct from that of capital; but in the United States, ownership in land and property became a very important form of investment from the eighteenth century onward. Under these conditions the "land and property interest" is simply reduced to a faction of capital (usually the money capitalists and the rentiers) investing in the appropriation of rent. This brings us to consider the important role of property companies, developers, banks, and other financial intermediaries (insurance companies, pension funds, savings and loan associations, etc.) in the land and property market. And I should also add that "homeownership" does not quite mean what it says, because most homeowners actually share equity with a financial institution and do not possess title to the property. In the United States, therefore, we have to think of the land and property interest primarily as a faction of capital investing in rental appropriation.

I shall assume for purposes of analytic convenience that a clear distinction exists between these classes and factions and that each pursues its own interests single-mindedly. In a capitalist society, of course, the whole structure of social relations is founded on the domination of capitalists over laborers. To put it this way is simply to acknowledge that the capitalists make the investment decisions, create the jobs and the commodities, and function as the catalytic agents in capitalist growth. We cannot hold, on the one hand, that America was created by the efforts of private entrepreneurs and deny, on the other, that capital dominates labor. Labor is not passive, of course, but its actions are defensive and at best confined to gaining a reasonable share of the national product. But if labor controlled the investment decisions, then we would not be justified any longer in describing our society as capitalist. Our interest here is not so much to focus on this primary antagonism but to examine the myriad secondary forms of conflict which can spin off from it to weave a complex web of arguments over the production and use of the built environment. Appropriators (landlords and property owners) may be in conflict with construction interests, capitalists may be dissatisfied with the activities of both factions, and labor may be at odds with all of the others. And if the transport system or the sewage system does not work, then both labor and capital will be equally put out. Let us consider two examples that, in spite of their hypothetical nature, illustrate the complex alliances that can form and shed some light on the kinds of problems which urban planners typically face.

I start with the proposition that the price of existing resources in the built environment – and, hence, the rate of rental appropriation – is highly sensitive to the costs and rate of new construction. Suppose the construction interests are badly organized, in a slump, or unable to gain easy access to cheap land and that the rate of new construction is low and the costs high. Under these conditions, those seeking the appropriation of rent possess the power to increase their rate of return by raising rental on, say, housing. Labor may resist; tenant organizations may spring up and seek to control the rate of rental appropriation and to keep the cost of living down. If they succeed, tenant organizations may even drive the rate of return on existing resources downward to the point where investment withdraws entirely (perhaps producing abandonment). If labor lacks organization and power in the community but is well organized and powerful in the workplace, a rising rate of appropriation may result in the pursuit of higher wages, which, if granted, may lower the rate of profit and accumulation. A rational response of the capitalist class under these conditions is to seek an alliance with labor to curb excessive rental apropriations, to free land for new construction, and to see to it that cheap (perhaps even subsidized) housing is built for the laboring class. We can see this sort of coalition in action when large corporate interests in suburban locations join with civil rights groups in trying to break suburban zoning restrictions that exclude low-wage populations from the suburbs. An exploration of this dimension of conflict can tell us much about the structure of contemporary urban problems.

The second case I shall consider arises out of the general dynamic of capitalist accumulation, which, from time to time, generates chronic overproduction, surplus real productive capacity, and idle money capital desperately in need of productive outlets. In such a situation, money is easily come by to produce long-term investments in the built environment, and a vast investment wave flows into the production of the built environment, which serves as a vent for surplus capital – such was the boom experienced from 1970 to 1973. But at some point the existence of overproduction becomes plain to see – be it office space in Manhattan or housing in Detroit – and the property boom collapses in a wave of bankruptcies and "refinancings" (consider, for example, the fall of the secondary banks associated with the London property market in 1973 and the dismal performance of the Real Estate Investment Trusts in the United States with some $11 billion in assets, half of which are currently earning no rate of return at all). What becomes evident in this case is that excessive investment brings in its wake disinvestment and devaluation of capital for at least some segment of the landed interest. The construction interest is also faced with an extremely difficult pattern of booms and slumps which militates against the creation of a viable long-term organization for the coherent production of the built environment. If labor sinks part of its equity into the property market, then it, too, may find its savings devalued by such processes; and, through community organization and political action, it may seek to protect itself as well as it can. In this case also, we can discern a structure to our urban problems which is explicable in terms of the conflicting requirements of the various classes and factions as they face up to the problems created by the use of the built environment as a vent for surplus capital in a period of overaccumulation (see Chap. 1).

These dimensions of conflict are cut across, however, by a completely different set of considerations which arise out of the fact that the built environment is composed of assets that are typically both long-lived and fixed in space. This means that we are dealing with commodities that must be produced and used under conditions of "natural monopoly" in space. It also happens that, since the built environment is to be conceived of as a complex composite commodity, the individual elements have strong "externality" effects on other elements. We thus find that competition for use of resources is monopolistic competition in space – that capitalists can compete with capitalists for advantageously positioned resources, that laborers can compete with laborers for survival chances, access, and the like, and that land and property owners seek to influence the positioning of new elements in the built environment (particularly transport facilities) so as to gain indirect benefits. The basic structure of class and factional conflicts is therefore modified and in some instances totally transformed into a structure of geographical conflict which pits laborers in the suburbs against laborers in the city, capitalists in the industrial Northeast against capitalists in the Sun Belt, and so on.

The distinctive role and tasks of the planner have to be understood against the background of the strong currents of both interclass and factional conflict, on the one hand, and the geographical competition that natural monopolies in space inevitably generate, on the other.

III. THE PRODUCTION, MAINTENANCE, AND MANAGEMENT OF THE BUILT ENVIRONMENT

The built environment must incorporate the necessary use values to facilitate social reproduction and growth. Its overall efficiency and rationality can be tested and measured in terms of how well it functions in relationship to these tasks. The sophisticated model builders within the planning fraternity have long sought to translate this conception into a search for some idealized *optimum optimorum* for the city or for regional structure. Such a search can be entertaining and can generate insights into certain typical characteristics of urban structure, but as an enterprise it is utopian, idealized, and fruitless. A more down-to-earth analysis suggests that the indications of failure of the built enviroment to provide the necessary use values are not too hard to spot. The evidence of crisis and of failure to reproduce effectively or to grow at a steady rate of accumulation is a clear indicator of a lack of balance which requires some kind of corrective action.

Unfortunately, "crisis" is a much overused word. Anybody who wants anything in this society is forced to shout "crisis" as loudly as possible in order to get anything done. For the underprivileged and the poor, the crisis is permanent and endemic. I shall take a narrower view and define a crisis as a particular conjuncture in which the reproduction of capitalist society is in jeopardy. The main signals are falling rates of profit; soaring unemployment and inflation; idle productive capacity and idle money capital lacking profitable employment; and financial, institutional, and political chaos and civil strife. And we can identify three wellsprings out of which crises in capitalist society typically flow. First, an imbalanced outcome of the struggle between the classes or factions of classes may permit one class or faction to acquire excessive power and so destabilize the system (well-organized workers can force the wage rate up and the accumulation rate down; finance capital may dominate all other factions of capital and engage in uncontrolled speculative binges, and so forth). Second, accumulation pushes growth beyond the capacity of the sustaining natural resource base at the same time as technological innovation slackens. Third, a tendency toward overaccumulation and overproduction is omnipresent in capitalist societies because individual entrepreneurs, pursuing their own individual self-interest, collectively push the dynamic of aggregative accumulation away from a balanced growth path (see Harvey 1982, chap. 7).

The particular role of the built environment in all of this is complex in its details but simple in principle (see Chap. 1). Failure to invest in those elements in the built environment which contribute to accumulation is no different in principle than the failure of entrepreneurs to invest and reinvest in fixed capital equipment. The problem with the built environment is that much of it functions as collective fixed capital (transport, sewage, and disposal systems, etc.). Some way has to be found, therefore, to ensure a flow of investment into the built environment and to ensure that individual investment decisions are coordinated in both time and space so that the aggregative needs of capitalist producers are met. By the same token, failure to invest in the means of consumption for labor may raise the wage rate, generate civil strife, or (in the worst kind of eventuality) physically diminish the supply of labor. In both cases, failure to invest in the right quantities, at the right times, and in the right places can be a progenitor of a crisis of accumulation and growth. Overinvestment in the built environment is, in contrast, simply a devaluation of capital which nobody, surely, welcomes. And so we arrive at the general conception of the *potential* for a harmonious, balanced investment process in the built environment. Any departure from this path will entail either underinvestment (and a constraint upon accumulation) or overinvestment (and the devaluation of capital). The problem is to find some way to ensure that such a potentiality for balanced growth is realized under the conditions of a capitalist investment process.

The built environment is long-lived; fixed in space; and a complex, composite commodity, the individual elements of which may be produced, maintained, managed, and owned by quite diverse interests. Plainly, there is a problem of coordination, because mistakes are very difficult to recoup and individual producers may not always act to produce the proper mix of elements in space. The time stream of benefits to be derived also poses some peculiar problems. The physical landscape created at one point in time may be suited to the needs of society at that point but become antagonistic later as the dynamics of accumulation and societal growth alter the use value requirements of both capital and labor. Tensions may then arise because the long-lived use values embedded in the built environment cannot easily be altered on a grand scale – witness the problems endemic to many of the older industrial and commercial cities in the industrial Northeast of the United States at the present time.

Investment in the built environment can be coordinated with general social requirements in one way, or in a mix of three:

1. Allocations can be arrived at through market mechanisms. Elements that can be privately appropriated under the legal relations of private property – houses, factories, offices, stores, warehouses, etc. – can be rented and traded. This sets up the price signals that, under pure competitive bidding, will allocate land and plant to the best-paying uses. The price signals also make it possible to calculate a rate of return on new investments, which usually generates a flow of new investment to wherever the rate of return is above that to be had, given similar risks, in other sectors of the capital market. But the innumerable externality effects and the importance of "public good" items that cannot be privately appropriated – streets, sidewalks, etc. – generate frequent market failures and imperfections so that in no country is investment in the built environment left entirely to competitive market mechanisms.

2. Allocations may be arrived at under the auspices of some hegemonic controlling interest – a land or developer monopoly, controlling financial

interests, and the like.[3] This is not an irrational move, because a large-scale enterprise coordinating investments of many different types can "internalize the externalities" and thereby make more rational decisions from the investor standpoint – the land grant railroads provide an excellent historical example of such monopolistic control, while Rouse's Columbia provides a contemporary example. The trouble with monopolization and hegemonic control is that the pricing system becomes artificial (and this can lead to misallocations), while there is nothing to ensure that monopoly power is not abused.

3. State intervention is an omnipresent feature in the production, maintenance, and management of the built environment.[4] The transport system – prime example of a "natural monopoly" in space – has always posed the problem of private gain versus public social benefit, private property rights versus aggregative social needs. The abuse of monopoly power (which it is all too easy to accumulate in spatial terms) has ever brought forth state regulation as a response. The pervasive externality effects have in all countries led to state regulation of the spatial order to reduce the risks that attach to long-term investment decisions. And the "public goods" elements in the built environment – the streets, sidewalks, sewer and drainage systems, etc. – which cannot feasibly be privately appropriated have always been created by direct investment on the part of the agencies of the state. The theme of "public improvement" has been writ large in the history of all American cities.

The exact mix of private market, monopolistic control, and state intervention and provision has varied with time as well as from place to place. Which mix is chosen or, more likely, arrived at by a complex historical process is not that important. What is important is that it should ensure the creation of a built environment that serves the purpose of social reproduction and that it should do so in such a manner that crises are avoided as far as is possible.

IV. URBAN PLANNING AS PART OF THE INSTRUMENTALITIES OF STATE

The proper conception of the role of the state in capitalist society is controversial.[5] I shall simply take the view that state institutions and the

[3] Some idea of the extent of hegemonic control exercised by finance capital over the land and property market can be gained from L. Downie (1974); G. Barker, J. Penney, and W. Seccombe (1973); and P. Ambrose and R. Colenutt (1975). See also Chapter 3.

[4] The French urbanists have worked on this aspect most carefully as in M. Castells and F. Godard (1973) and C. G. Pickvance (1976). See also the various essays in *Antipode* 7, no. 4 – a special issue entitled *The Political Economy of Urbanism* – and D. Harvey (1975).

[5] See, for example, E. Altvater (1973, 1:96–108 and 3:76–83); R. Miliband (1968); N. Poulantzas (1973); and J. O'Connor (1973).

processes whereby state powers are exercised must be so fashioned that they, too, contribute, insofar as they can, to the reproduction and growth of the social system. Under this conception we can derive certain basic functions of the capitalist state. It should

1. help to stabilize an otherwise rather erratic economic and social system by acting as a "crisis-manager";
2. strive to create the conditions for "balanced growth" and a smooth process of accumulation;
3. contain civil strife and factional struggles by repression (police power), cooptation (buying off politically or economically), or integration (trying to harmonize the demands of warring classes or factions).

The state can effectively perform all of these functions only if it succeeds in internalizing within its processes the conflicting interests of classes, factions, diverse geographical groupings, and so on. A state that is entirely controlled by one and only one faction, that can operate only repressively and never through integration or cooptation, will likely be unstable and will likely survive only under conditions that are, in any case, chronically unstable. The social democratic state is one that can internalize diverse conflicting interests and that, by means of the checks and balances it contains, can prevent any one faction or class from seizing direct control of all of the instrumentalities of government and putting them to its own direct use. Yet the social democratic state is still a capitalist state in the sense that it is a capitalist social order that it is helping to reproduce. If the instrumentalities of state power are turned against the existing social order, then we see a crisis of the state, the outcome of which will determine whether the social order changes or whether the organization of the state reverts to its basic role of serving societal reproduction.

The urban planner occupies just one niche within the total complex of the instrumentalities of state power. The internalization of conflicting interests and needs within the state typically puts one branch of the bureaucracy at loggerheads with another, one level or branch of government against another, and even different departments at odds with one another within the same bureaucracy. In what follows, however, I shall lay aside these diverse cross-currents of conflict and seek to abstract some sense of the real limitations placed upon the urban planner by virtue of his or her role and thereby come to identify more clearly the nature of the role itself. To hasten the argument along, I shall simply suggest that the planner's task is to contribute to the processes of social reproduction and that in so doing the planner is equipped with powers vis-à-vis the production, maintenance, and management of the built environment which permit him or her to intervene in order to stabilize, to create the conditions for "balanced growth," to contain civil strife and factional struggles by repression, cooptation, or integration. And to fulfill these goals successfully, the planning process as a whole (in which the planner fulfills only one set of tasks) must be relatively open. This conception may appear unduly simplistic, but a down-to-earth analysis of what planners actually do, as opposed to what they or the mandarins of the planning fraternity think they do, suggests that the conception is not far from the mark. And the history of those who seek to depart radically from this fairly circumscribed path suggests that they either encounter frustration or else give up the role of planner entirely.[6]

V. THE PLANNER'S KNOWLEDGE AND IMPLIED WORLD VIEW

In order to perform the necessary tasks effectively, the planner needs to acquire an understanding of how the built environment works in relationship to social reproduction and how the various facets of competitive, monopolistic, and state production of the built environment relate to one another in the context of often conflicting class and factional requirements. Planners are therefore taught to appreciate how everything relates to everything else in an urban system, to think in terms of costs and benefits (although they may not necessarily resort to techniques of cost-benefit analysis), and to have some sympathetic understanding of the problems that face the private producers of the built environment, the landlord interest, the urban poor, the managers of financial institutions, the downtown business interests, and so on. The accumulation of planning knowledge arises through incremental understandings of what would be the "best" configuration of investment (both spatial and in terms of quantitative balance) to facilitate social reproduction. But the most important shifts in understanding come in the course of those crises in which something obviously must be done because social reproduction is in jeopardy.

The planner requires something else as well as a basic understanding of how the system works from a purely technical standpoint. In resorting to tools of repression, cooptation, and integration, the planner requires justification and legitimation, a set of powerful arguments with which to confront warring factional interests and class antagonisms. In striving to affect reconciliation, the planner must perforce resort to the idea of the potentiality for harmonious balance in society. And it is on this fundamental notion of social harmony that the ideology of planning is built. The planner seeks to intervene to restore "balance," but the "balance" implied is that which is

[6] A good example of how planners might move down such a path is written up in R. Goodman (1971).

necessary to reduce civil strife and to maintain the requisite conditions for the steady accumulation of capital. From time to time, of course, planners may be captured (by corruption, political patronage, or even radical arguments) by one class/faction or another and thereby lose the capacity to act as stabilizers and harmonizers – but such a condition, though endemic, is inherently unstable, and the inevitable reform movement will most probably sweep it away when it is no longer consistent with the requirements of the social order. The role of the planner, then, ultimately derives its justification and legitimacy from intervening to restore that balance which perpetuates the existing social order. And the planner fashions an ideology appropriate to the role.

This does not necessarily mean that the planner is a mere defender of the status quo. The dynamics of accumulation and of societal growth are such as to create endemic tensions between the built environment as is and as it should be, while the evils that stem from the abuse of spatial monopoly can quickly become widespread and dangerous for social reproduction. Part of the planner's task is to spot both present and future dangers and to head off, if possible, an incipient crisis of the built environment. In fact, the whole tradition of planning is progressive in the sense that the planner's commitment to the ideology of social harmony – unless it is perverted or corrupted in some way – always puts the planner in the role of "righter of wrongs," "corrector of imbalances," and "defender of the public interest." The limits of this progressive stance are clearly set, however, by the fact that the definitions of the public interest, of imbalance, and of inequity are set according to the requirements for the reproduction of the social order, which is, whether we like the term or not, a distinctively capitalistic social order.

The planner's knowledge of the world cannot be separated from this necessary ideological commitment. Existing and planned orderings of the built environment are evaluated against some notion of a "rational" socio-spatial ordering. But it is the capitalistic definition of rationality to which we appeal (Godelier 1972). The principle of rationality is an ideal – the central core of a pervasive ideology – which itself depends upon the notion of harmonious processes of social reproduction under capitalism. The limits of the planner's understanding of the world are set by this underlying ideological commitment. In the reverse direction, the planner's knowledge is used ideologically, as both legitimation and justification for certain forms of action. Political struggles and arguments may, under the planner's influence, be reduced to technical arguments for which a "rational" solution can easily be found. Those who do not accept such a solution are then open to attack as "unreasonable" and "irrational." In this manner both the real understanding of the world and the prevailing ideology fuse into a world view. I do not mean to imply that all planners subscribe to the same world view – they manifestly do not, and it would be dysfunctional were they to do so. Some planners are very technocratic and seek to translate all political issues into technical problems, while other planners take a much more political stance. But whatever their position the fusion of technical understandings with a necessary ideology produces a complex mix within the planning fraternity of capacity to understand and to intervene in a realistic and advantageous way and capacity to repress, coopt, and integrate in a way that appears justifiable and legitimate.

VI. CIVIL STRIFE, CRISIS OF ACCUMULATION, AND SHIFTS IN THE PLANNER'S WORLD VIEW

The planner's world view, defined as the necessary knowledge for appropriate intervention and the necessary ideology to justify and legitimate action, has altered with changing circumstances. But knowledge and ideology do not change overnight. The concepts, categories, relationships, and images through which we interpret the world are, so to speak, the fixed capital of our intellectual world and are no more easily transformed than the physical infrastructures of the city itself. It usually takes a crisis, a rush of ideas pouring forth under the pressure of events, radically to change the planner's world view, and even then radical change comes but slowly. And while the fundamentals of ideology – the notion of social harmony – may stay intact, the meanings attached must change according to whatever it is that is out of balance. The history of capitalist societies these last two hundred years suggests, however, that certain problems are endemic, problems that simply will not go away no matter how hard we try. Consequently we find that the shifting world view of the planner exhibits an accumulation of technical understandings combined with a mere swaying from side to side in ideological stance from which the planner appears to learn little or nothing. Let me illustrate.

The capitalist growth process has been punctuated, at quite regular intervals, by phases of acute social tension and civil strife. These phases are not historical accidents but can be traced back to the fundamental characteristics of capitalist societies and the growth processes entailed. I have not space to elaborate on this theme here, but it is important to note that the organization of work under capitalism is predicated on a separation between "working" and "living," on control of work by the capitalist and alienated labor for the employee, and on a dynamic relation between the wage rate and the rate of profit which is founded on the social necessity for a surplus of labor which may vary quantitatively according to time and place. Generally speaking, it is the concentration of low-wage populations and unemployment in either time or space that sets the stage for civil strife.

The response is some mixture of repression, cooptation, and integration. The urban planner's role in all of this is to define policies that facilitate social control and that serve to reestablish social harmony through cooptation and integration. Consider, for example, the spatial distribution of the population, particularly of the unemployed and low-wage earners. The revolutions of 1848, the Paris Commune of 1871, the urban violence that accompanied the great railroad strikes of 1877 in the United States, and Chicago's celebrated "Haymarket affair" of 1886 demonstrated the revolutionary dangers associated with high concentrations of what Charles Loring Brace (1889) called, in the 1870s, the "dangerous classes" of society. The problem could be dealt with by a policy of dispersal, which meant that ways had to be found to permit the poor and the unemployed to escape their chronic entrapment in space. The urban working class had to be dispersed and subjected to what reformers on both sides of the Atlantic called "the moral influence" of the suburbs (Tarr 1973). Suburbanization facilitated by cheap communication was seen as part of the solution. The urban planners and reformers of the time pressed hard for policies of dispersal via mass transit facilities such as those provided under the Cheap Trains Act of 1882 in Britain and the streetcars in the United States, while the search for cheap housing and means to promote social stability through working-class homeownership began in earnest. In much the same way, planners in the 1960s responded to the urban riots by seeking ways to disperse the ghetto by improved transport relations, promoting homeownership, and opening up housing opportunities in the suburbs (although this time round, the Victorian rhetoric of "moral influence" was replaced by the more "rational" appeal of "social stability"). In the process, the laboring classes undoubtedly gain in real living standards while the planner acts as advocate for the poor and the underprivileged, raises the cry of social justice and equity, expresses moral outrage at the conditions of life of the urban poor, and reaches for ways to restore social harmony.

The alternative to dispersal is what we now call "gilding the ghetto," and this, too, is a well-tried tactic in the struggle to control civil strife in urban areas. As early as 1812, the Reverend Thomas Chalmers (1821–26) raised the specter of a tide of revolutionary violence sweeping Britain as working-class populations steadily concentrated in large urban areas. Chalmers saw "the principle of community" as the main bulwark of defense against this revolutionary tide – a principle that sought to establish harmony between the classes around the basic institutions of community. The principle also entailed a commitment to community improvement, the attempt to instill some sense of civic or community pride capable of transgressing class boundaries. The church was then the most important institution, but we now think of other instrumentalities also – political inclusion, citizen participation, and community commitment to educational, recreational, and other services as well as the sense of pride in neighborhood which inevitably means a "better" quality of built environment. From Chalmers through Octavia Hill and Jane Addams, to Model Cities and citizen participation, we have a continuous thread of an argument which suggests that social stability can be restored in periods of social unrest by an active pursuit of "the principle of community" and all that this means in the way of community betterment and social improvement – and again, the planner typically acts as advocate, as catalyst in promoting the spirit of community improvement.

One dimension of this idea of "improvement" is that of environmental quality. Olmstead was perhaps the first fully to recognize that the efficiency of labor might be enhanced by providing a compensatory sense of harmony with nature in the living place, although it is important to recognize that Olmstead was building on a rather older tradition.[7] At issue here is the relation to nature in a most fundamental sense. Industrial capitalism, armed with the factory system, organized a work process that transformed the relation between the worker and nature into a travesty of its older artisan self. Reduced to a thing, a commodity, a mere "factor" of production, the worker became alienated from the product of work, the process of production, and ultimately from nature itself, particularly in the industrial city, where, as Dickens (1961) puts it, "Nature was as strongly bricked out as killing airs and gases were bricked in." The romantic reaction against the new industrial order ultimately led in the practice of urban design and planning to the attempt to counter in the sphere of consumption for what had been lost in the sphere of production. The attempt to "bring nature back into the living environments" within the city has been a consistent theme in planning since Olmstead's time. Yet it is, in the final analysis, an attempt based in what Raymond Williams (1973) calls "an effective and imposing mystification," for there is something in the relation to nature in the work process that can never be compensated for in the consumption sphere. The planner, armed with concepts of ecological balance and the notion of harmony with nature, acts once more as advocate and brings real gains. But the real solutions to these problems lie elsewhere, in the work process itself.

Civil strife and social discontent provide only one set of problems which the planner must address. The dynamic of accumulation, with its periodic crises of overaccumulation, poses an entirely different set. The crises are not accidental. They are to be viewed, rather, as major periods of "rationalization," of "shake-outs and shake-downs" that restore balance to an economic system temporarily gone mad. The fact that crises perform this rationalizing function is no comfort to those caught in their midst. And at such conjunctures planners must either simply administer the budget cuts and plan the shrinkage according to the strict requirements of an externally

[7] See, for example, T. Bender (1975) and R. A. Walker (1976).

imposed fiscal logic of the sort recently applied to New York or seek to head a movement for a forced rationalization of the urban system. The pursuit of the city beautiful is replaced by the search for the city efficient; the cry of social justice is replaced by the slogan "efficiency in government"; and those planners armed with a ruthless cost-benefit calculus, a rational and technocratic commitment to efficiency for efficiency's sake, come into their own.

"Rationalization" means, of course, doing whatever must be done to reestablish the conditions for a positive rate of accumulation. When economic growth goes negative – as it did, for example, in 1893, in the early 1930s, or in 1970 and 1974, then the reproduction of the social order is plainly in doubt. The task at such conjunctures is to find out what is wrong and to right it. The physical infrastructure of the city may be congested, inefficient, and too costly to use for purposes of production and exchange. Such barriers (which were obvious to all in the Progressive Era, for example) must be removed, and if the planner does not willingly help to do so, then the escalating competition between jurisdictions for "development" at times of general decline will force the planner into action if he or she values the tax base (this kind of competitive pressure often leads communities to subsidize profits). If the problem lies in the consumption sphere – underconsumption or erratic movements in aggregate personal behaviors – then the state may seek to manage consumption either by fiscal devices or through collectivization. The management of collective consumption by means of the built environment at such points becomes a crucial part of the planner's task.[8] If the problem lies in lack or excess of investment in the built environment, then the planner must perforce set to work to stimulate investment or to manage and rationalize devaluation with techniques of "planned shrinkage," urban renewal, and even the production of "planning blight" (which amounts to nothing more than earmarking certain areas for devaluation).

I list these various possibilities because it is not always self-evident as to what must be done in the heat of a crisis of accumulation. At such conjunctures our knowledge of the system and how it works is crucial for action, unless we are to be led dangerously near the precipice of cataclysmic depression. And it is exactly at such points that the world view of the planner, restricted as it is by an ideological commitment, appears most defective, while the ideological stance of the planner may have to shift under the pressure of events from advocacy for the urban poor to one dedicated to business rationality and efficiency in government.

But ideologies, I have argued, do not change that easily, nor does our knowledge of the world. And so we find, at each of the major turning points

[8] Again, the French urbanists have discussed this idea at length in, for example, E. Preteceille (1975) and M. Castells (1975).

in our history, a *crisis of ideology.*[9] Past commitments must obviously be abandoned because they hinder our power to understand and most certainly lose their power to legitimate and justify (imagine trying to justify what happened to New York City's budget in the mid 1970s by appealing to concepts of social justice). And as the pillars of the planner's world view slowly crumble, so the search begins for a new scaffolding for the future. At such a juncture, it becomes necessary to plan the ideology of planning.

VII. PLANNING THE IDEOLOGY OF PLANNING

By the mid-1970s it became clear that the planning inspirations of the 1960s had faded and that our main task was to define new horizons for planning into the 1980s – new technologies, new instrumentalities, new goals . . . new everything in fact *except* a new ideology. Yet if my analysis is correct, the real task was to plan the ideology of planning to fit the new economic realities rather than to meet the social unrest and civil strife of the 1960s.

Since many of those who inspired us in the 1960s are still active, it is useful to ask what, if anything, went wrong. The crucial problem of the 1960s was civil strife and in particular the concentrated form of it associated with the urban riots in central city areas. That strife had to be contained by repression, cooptation, and integration. In this the planner, armed with diverse ideologies and a variety of world views, played a crucial role. The dissidents were encouraged to go through "channels," to adhere to "procedures laid down," and somewhere down that path the planner lay in wait with a seemingly sophisticated technology and an intricate understanding of the world, through which political questions could be translated into technical questions that the mass of the population found hard to understand. But discontent cannot so easily be controlled, and so the other string to the planner's bow was to find ways to disperse the urban poor, to divide and control them and to ameliorate their conditions of life (Piven and Cloward 1971; Cloward and Piven 1974). The management of this process fell very much within the domain of urban planning, and it generated conflicting ideological stances and world views within the planning profession itself. At first sight (and indeed at the time) it seemed as if planning theory was fragmented in the 1960s as different segments of the planning fraternity moved according to their position or inclination to one or the other pole of the ideological spectrum. With the benefit of hindsight, we can see that this

[9] There is an important connection between crises in ideology and legitimation – see, for example, J. Habermas (1975); for a history of shifting ideology in urban development, see R. A. Walker (1976).

process was nothing more than the internalization within the planning apparatus of conflicting social pressures and positions. And this internalization and the oppositions that it provoked proved functional, no matter what individuals thought or did. The technocrats helped to define the outer bounds of what could be done at the same time as they sought for new instrumentalities to accomplish dispersal and to establish social control. The advocates for the urban poor and the instrumentalities that they devised provided the channels for cooptation and integration at the same time as they pushed the system to provide whatever could be provided, being careful to stop short at the boundaries that the technocrats and "fiscal conservatives" helped to define. Those who pushed advocacy too far were either forced out or deserted planning altogether and became activists and political organizers.

Judged in terms of their own ideological rhetoric the pursuers of social justice failed, much as they did in the Progressive Era, to accomplish what they set out to do, although the position of the "dangerous classes" in society undoubtedly improved somewhat in the late 1960s. But judged in relation to the reduction of civil strife, the reestablishment of social control, and the "saving" of the capitalist social order, the planning techniques and ideologies of the 1960s were highly successful. Those who inspired us in the 1960s can congratulate themselves on a job well done.

But conditions changed quite radically in 1969–70. Stagflation emerged as the most serious problem, and the negative growth rate of 1970 indicated that the fundamental processes of accumulation were in deep trouble. A loose monetary policy – the most potent tool in the management of the "political business cycle" – saw us through the election of 1972. But the boom was speculative and heavily dependent upon a massive overinvestment in the land, property, and construction sectors that easy money typically encourages. By the end of 1973 it was plain that the built environment could absorb no more in the way of surplus capital, and the rapid decline in property and construction, together with financial instability, triggered the subsequent depression. Unemployment doubled, real wages began to move downward under the impact of severe "labor-disciplining" policies, social programs began to be savagely cut, and all of the gains made after a decade of struggle in the 1960s by the poor and underprivileged were rolled back almost within the space of a year. The underlying logic of capitalist accumulation asserted itself in the form of a crisis in which real wages diminished in order that inflation be stabilized and appropriate conditions for accumulation be reestablished.

The pressure from this underlying logic was felt in all spheres. Local budgets had to shift toward fiscal conservatism and had to alter priorities from social programs to programs to stimulate and encourage development (often by subsidies and tax benefits). Planners talked grimly of the "hard, tough decisions" that lay ahead. Those that sought social justice as an end in itself in the 1960s gradually shifted their ground as they began to argue that social justice could best be achieved by ensuring efficiency in government. Those that sought ecological balance and conservation in its own right in the 1960s began to appeal to principles of rational and efficient management of our resources. The technocrats began the search for ways to define more rational patterns of investment in the built environment, to calculate costs and benefits more finely than ever. The gospel of efficiency came to reign supreme.

All of this presupposes the capacity to accomplish a transformation of ideological balance within the planning fraternity – a transformation that turns out to be almost identical to that which was successfully accomplished during the Progressive Era. It can, of course, be done. But it takes effort and fairly sophisticated argument to do it. And the transformation is made that much easier because the fundamentals of ideology remain intact. The commitment to the ideology of harmony within the capitalist social order remains the still point upon which the gyrations of planning ideology turn.

But if we step aside and reflect awhile upon the tortuous twists and turns in our history, a shadow of doubt might cross our minds. Perhaps the most imposing and effective mystification of all lies in the presupposition of harmony at the still point of the turning capitalist world. Perhaps there lies at the fulcrum of capitalist history not harmony but a social relation of domination of capital over labor. And if we pursue this possibility, we might come to understand why the planner seems doomed to a life of perpetual frustration, why the high-sounding ideals of planning theory are so frequently translated into grubby practices on the ground, how the shifts in world view and in ideological stance are social products rather than freely chosen. And we might even come to see that it is the commitment to an alien ideology which chains our thought and understanding in order to legitimate a social practice that preserves, in a deep sense, the domination of capital over labor. Should we reach *that* conclusion, then we would surely witness a markedly different reconstruction of the planner's world view than we are currently seeing. We might even begin to plan the reconstruction of society, instead of merely planning the ideology of planning.

8

The Urbanization of Capital

The language of any question has the awkward habit of containing the elements of its own response. For this reason I have always attached particular importance to Marx's comment that "frequently the only possible answer is a critique of the question and the only possible solution is to negate the question" (*Grundrisse,* 127). The eternal skepticism of Marxian scientific endeavor resides precisely in that methodological prescription.

The question I began with, more than a decade ago now, was roughly this: can we derive a theoretical and historical understanding of the urban process under capitalism out of a study of the supposed laws of motion of a capitalist mode of production? I quickly became convinced that the answer was yes, provided those laws could be specified more rigorously in terms of temporal and spatial dynamics. But was this the right question? A decade of thinking and writing on the subject points to a reformulation. I now ask, how does capital become urbanized, and what are the consequences of that urbanization? The answer to that question has, I submit, profound implications for understanding the future of capitalism as well as the prospects for transition to some alternative mode of production.

But let me begin with some remarks on how we might conceptualize urbanization in the context of a predominantly capitalist mode of production.

I. THE PRODUCTION OF URBANIZATION UNDER A CAPITALIST MODE OF PRODUCTION

The use values necessary to the reproduction of social life under capitalism are basically produced as commodities within a circulation process of capital that has the augmentation of exchange values as its primary goal. The standard form of this circulation process can be symbolized as follows:

$$M \rightarrow C \begin{cases} LP \\ MP \end{cases} \ldots P \ldots C' \rightarrow M + \triangle m \rightarrow \text{etc.}$$

where M stands for money; C and C' for commodities; MP for all kinds of means of production, including machinery, energy inputs, raw materials, and partially finished products; LP for labor power; P for production; and Δm for profit. This was the system that Marx scrutinized so thoroughly in order to establish its inherent laws of motion. He showed, among other things, that capitalism had to be both expansionary and technologically dynamic; that profit depended on the exploitation of living labor power in production; and that this defined the central class relation and line of class struggle between buyers (capitalists) and sellers (workers) of labor power as a commodity. He also showed that the necessary expansion ("accumulation for accumulation's sake, production for production's sake") often conflicted with the impulsion to revolutionize the productive forces under such a system of class relations. The system is therefore unstable, degenerating into periodic crises of overaccumulation, a condition in which surpluses of capital and labor power exist unused side by side. Overaccumulation leads to devaluation and destruction of both capital and labor power unless some way can be found profitably to absorb them (cf. Harvey 1982).

The study of urbanization under such a mode of production requires much closer attention to spatial and temporal dynamics than Marx was prepared to give (though, as we saw in Chap. 2, Marx was quite aware of these aspects of the process). We can begin on the path toward some kind of integration by scrutinizing the different moments of money, commodities, labor power (and its reproduction), and production within the circulation of capital and the transitions (metamorphoses, Marx called them) between one moment and the other. We immediately see that each moment has a different capacity for geographical mobility and that the transitions inevitably entail some kind of spatial movement.

Let us look more closely at the point of commodity exchange ($M \rightarrow C$ and $C' \rightarrow M + \Delta m$). The intervention of money in exchange, Marx comments, permits the separation of purchases and sales in space and time. But how much separation? The analysis of the circulation of capital cannot really proceed without some kind of answer to that question. The spatial and temporal horizons of exchange are evidently socially determined. Investments in new systems of transport and communications reduce spatial barriers and roll back the possible geographical boundaries of exchange relations. Revolutions in the credit system relax and roll back temporal constraints, making long-term investments both possible and compatible with other production systems of radically different turnover times (between, for example, the production of power stations, corn, and short-order meals). Nevertheless, the buying and selling of commodities (including the purchase of machinery and other intermediate goods) entails the loss of time and money in overcoming spatial separation. This means that commodity markets become articulated into distinctive geographical trading patterns in which the efficiency of coordinations in space and time is a vital consideration.

The details of this are horrendously complex. Though the movement of commodities is constrained by the cost and time of transportation, credit money now moves as fast and with as few spatial constraints as information. Furthermore, each commodity has a different potentiality for movement (given its weight, perishability, and value), while time and cost introduce two dimensions often very different from physical distance. The social and geographical division of labor is in part an adaptation to these possibilities as well as an outcome of the general sociotechnical conditions prevailing in production. But the general point remains: when looked at from the standpoint of exchange, the circulation of capital is a geographical movement in time. I shall later seek to show that the geographical structures of commodity markets are more than mere reflections of capital circulation and function as real determinants of capitalism's dynamic.

The buying and selling of labor power deserves special scrutiny. Unlike other commodities, labor power has to go home every night and reproduce itself before coming back to work the next morning. The limit on the working day implies some sort of limit on daily travel time. Daily labor markets are therefore confined within a given commuting range (see Chap. 6). The geographical boundaries are flexible; they depend on the length of the working day within the workplace, the time and cost of commuting (given the modes and techniques of mass movement), and the social conditions considered acceptable for the reproduction of labor power (usually a cultural achievement of class struggle). A prima facie case exists, therefore, for considering the urban process in terms of the form and functioning of geographically integrated labor markets within which daily substitutions of labor power against job opportunities are in principle possible. The history of the urbanization of capital is at least in part a history of its evolving labor market geography.

The labor market is perpetually in the course of modification. Vast capital investments are directed to achieve relatively minor increases in the range of daily commuting possibilities. In-migration and population growth augment the supply of labor power but also entail considerable and sometimes vast capital investments for housing, food, and care. Investment in skills is a long-drawn-out process and also often absorbs large quantities of resources. The aspirations and demands of the laborers, particularly when enhanced by labor scarcity or organized class struggle, affect the quantities and qualities of labor supply in very particular ways, thus affecting the prospects for both accumulation and sociotechnical change in production. The result is considerable differentiation between geographically distinct labor markets. That, too, is what much of capitalist urbanization reflects.

Consider, now, the moment of production. With the singular but important exception of transport and communications (see Chap. 2), labor processes are pinned down to a particular place for the length of the working period (the time taken to produce the finished commodity). But even the short-order cook who has a very short working period may make use of fixed capital equipment that has an economic lifetime of several years. And some of that fixed capital cannot be moved without being destroyed. Production cannot change location in the middle of a working period without destroying some of the capital engaged, while the relative immobility and economic lifetime of the fixed capital used also severely constrain geographical mobility. The ability to move also depends, however, upon the sociotechnical conditions of production. The general Marxist approach is to see the evolution of those sociotechnical conditions of production as an outcome of intercapitalist competition and class struggle supplemented by spillover needs and effects from one sector of industry to another. But here I shall have to introduce a fundamental modification of the general Marxian account (see Chap. 6). I insist that intercapitalist competition and class struggle spark spatial competition for command of favorable locations and that the choice of sociotechnical mix is in part a response to the particularities of geographical situation. Viewing things this way helps us get a better handle on relations between the social and spatial divisions of labor in society.

We see immediately, for example, that the sociotechnical forms of the labor process are not independent of the geographical possibilities within structured labor and commodity markets and vice versa. The splitting of production into many specializations permits much greater sensitivity to geographical variation, thus allowing capitalists to exploit the differentials and accumulate capital faster than before. Detail functions can also be split up over space under the planned control of the corporation. This applies not only to the separation of design, planning, production, and marketing functions but also to the elements of a complex production system which can be produced in many different locations throughout the world before assembly into the final product. Such geographical separations have major impacts upon trading patterns and become feasible only to the degree that integrated production schedules can be organized efficiently over space. The general result is an evident tension between the virtues of geographical concentration to minimize spatial separation (the assembly of detail functions within the factory or the agglomeration of many firms within one urban center) and geographical dispersal, which has the virtue of providing opportunities for further accumulation by exploiting particular geographical advantages (natural or created). How that tension is resolved has important implications for the shape and form of the urban system. But the latter, insofar as it is shaped to facilitate tight temporal coordinations of flows in space, affects the way in which the tension is played out. The benefits to be had from conjoining the social with the geographical division of labor can then be parlayed into accelerating accumulation.

Production is typically separated from consumption under capitalism by market exchange. This has enormous implications for urbanization and urban structure. Work spaces and times separate out from consumption spaces and times in ways unknown in an artisan or peasant culture. The moment of consumption, like that of production, stands to be further fragmented. Vacation, leisure, and entertainment places separate from spaces of daily reproduction, and even the latter fragment into the lunch counter near the office, the kitchen, the neighborhood drugstore or bar. The spatial division of consumption is as important to the urban process as is the spatial division of labor – the qualities of New York, Paris, and Rome as well as the internal organization of these and other cities could not be understood without consideration of such phenomena. This is, however, a theme that remains underexplored in Marxian theory, in part because of the tendency to focus exclusively on production because it is the hegemonic moment in the circulation of capital.

Consumption also has to be looked at from another standpoint. The circulation of capital when viewed in aggregate presupposes the continuous expansion of effective demand in order to realize in the marketplace the value created through production. Where, then, does the effective demand come from? (cf. Harvey 1982, chap. 3). There are three broad sources: workers purchase wage goods (depending upon their achieved standard of living), the bourgeoisie purchases necessities and luxuries, and capitalists purchase investment goods (machinery and plant) and intermediate products. Each one of these markets has its own particular qualities and geographical sensitivities. The spatial division of labor puts a premium on continuous flow so that tight temporal and spatial coordinations and cost-minimization are mandatory pressures. Final consumption, particularly of luxuries for the bourgeoisie, is much less sensitive on these scores, though in the case of the wage laborers much depends upon customary living standards (the real wage) and the cost matrix within which the social reproduction of labor power takes place. However, we must also bear in mind that final consumption entails the use of a certain amount of fixed capital equipment (a consumption fund, I called it in Chap. 1). To the degree that this is fixed and of long life (housing, for example), so the "mode of consumption" tends to become fixed quantitatively, qualitatively, and geographically. The spatial division of consumption entails relatively permanent structurations of social and physical spaces both within and between urban centers.

Consider, finally, the moment of money itself. Money takes many forms, from the tangible commodity gold to the vague imprecision of an open line of credit. Money also has the peculiar quality of concentrating in time and space a form of universal power that is an expression of the world market in historical time. Money represents the greatest concentration of social power in the midst of the greatest possible dispersal. It can be used to overcome the geographical limitations of commodity and labor markets and fashion ever-greater dispersal of the spatial division of labor and of consumption. It can also transcend all other limits of geographical concentration and allow the assembly of massive social power in a few hands in a few places. It can be deployed over long time-horizons (as state debt, stocks and shares, mortgages, and so forth) or pulled together at particular moments for particular purposes. As a higher form of social power, it can dominate not only ownership of other means of production but also space and time as sources of social power (see *Consciousness and the Urban Experience,* chap. 1). The holding and command of money confers tremendous social power. But under capitalism that power is contingent upon the continuous use of money as capital.

Money, finance, and credit form a hierarchically organized central nervous system regulating and controlling the circulation of capital as a whole and expressing a class interest, albeit through private action (see Harvey 1982, chaps. 9 and 10). Financial markets separate out from commodity and labor markets and acquire a certain autonomy vis-à-vis production. Urban centers can then become centers of coordination, decision-making, and control, usually within a hierarchically organized geographical structure.

Let me summarize. An inspection of the different moments and transitions within the circulation of capital indicates a geographical grounding of that process through the patterning of labor and commodity markets, of the spatial division of production and consumption (under sociotechnical conditions that are in part an adaptation to geographical variations), and of hierarchically organized systems of financial coordination. Capital flow presupposes tight temporal and spatial coordinations in the midst of increasing separation and fragmentation. It is impossible to imagine such a material process without the production of some kind of urbanization as a "rational landscape" within which the accumulation of capital can proceed. Capital accumulation and the production of urbanization go hand in hand.

This perspective deserves modification on two counts. Profit depends upon realizing the surplus value created in production within a certain time. The turnover time of capital (the time taken to get back the initial outlay plus a profit) is a very important magnitude – hence derives the old adage "time is money." Competition produces strong pressures to accelerate turnover time. That same pressure has a spatial manifestation. Since movement across space takes time and money, competition forces capitalism toward the elimination of spatial barriers and "the annihilation of space by time" (see Chap. 2).

Building a capacity for increased efficiency of coordination in space and time is one of the hallmarks of capitalist urbanization. Considerations derived from a study of the circulation of capital dictate, then, that the urban matrix and the "rational landscape" for accumulation be subject to continuous transformations. In this sense also, capital accumulation, technological innovation and capitalist urbanization have to go together.

II. CITIES, SURPLUSES, AND THE URBAN ORIGINS OF CAPITALISM

The connection between city formation and the production, appropriation, and concentration of an economic surplus has long been noted (see Harvey 1973, 216–26). The circulation of capital also presupposes the prior existence of surpluses of both capital and labor power. But closer inspection of its dynamic shows that capital circulation, once set in motion, produces capital surpluses (in the form of profit) coupled with relative labor surpluses produced through labor-saving revolutions in the sociotechnical conditions of production. Much of the history of capitalism can be written around this theme of the production and absorption of capital and labor surpluses. Strong phases of balanced and seemingly self-sustained growth occur when capitalism produces exactly those surpluses it needs in order to continue on its expansionary path. But the tendency toward overaccumulation poses the problem of how to absorb or dispose of the surpluses without the devaluation or destruction of both capital and labor power. This tension between the need to produce and to absorb surpluses of both capital and labor power lies at the root of capitalism's dynamic. It also provides a powerful link to the history of capitalist urbanization. I shall, in what follows, make that link the pivot of much of my analysis.

In the early stages of capitalism, the surpluses were largely produced by processes external to the direct circulation of capital. The violent expropriation of the means of production through primitive accumulation or more subtle maneuvers of appropriation put capital surpluses in the hands of the few while the many were forced to become wage laborers in order to live. Capital exists potentially in many forms, however, so it was the various moves of appropriation of money, goods, productive assets (land, built environments, means of communication, and so forth), and rights to labor power and the conversion of all of these into commodities with exchange value that really counted. The appropriation, mobilization, and geographical concentration of these surpluses of capital and labor power in commodity form was a vital moment in capitalism's history in which urbanization played a key role (cf. Braudel 1984). The urban concentration of wealth by merchants (looting the world of both money and commodities through unfairly or badly structured exchange); the transformation of landed property into a commodity for the production of urban-based wealth through direct monetization or subversion by usurers; and the direct extraction of surpluses from the countryside through money rents, state taxation, and other mechanisms of redistribution (such as that organized through the church) were some of the means whereby surplus capital was mobilized and geographically concentrated in a few hands. The use of these surpluses to build physical infrastructures, communication systems, and market centers formed a potential basis for capital circulation at the same time as the assembly of commodity use values (including wage laborers) in the urban centers created the prior conditions under which the circulation of capital could be more easily launched.

Urbanization, together with money rent, usurers' interest, merchants' profit, and state taxation, had to appear on the historical stage before the standard form of circulation of capital through production could begin (cf. *Capital* 1:165). The historical sequence was exactly the reverse, therefore, of the analytical and logical sequence we would now use to analyze the relations of production and distribution and of long-term investment in physical and social infrastructures in their urban context (see Chap. 1 and fig. 3). A built environment potentially supportive of capitalist production, consumption, and exchange had to be created before capitalism won direct control over immediate production and consumption. Social infrastructures for the control of civil society, most particularly with respect to labor markets, also had to be put in place before capital accumulation through production could freely develop. The political power and authority of the state had to be deployed in ways favorable to primitive accumulation and the mobilization of capital and labor surpluses before the material base existed for capitalist domination of the state or even for the formation of some urban-based class alliance in which capitalists had an important role. The rise of urban centers with a ruling class acquisitive of wealth and specie, mercantilist in philosophy, and possessed of superior authority and military power was, Braudel (1984) shows, a crucial moment in the rise of capitalism. The maturation of capitalism rested on a process of gradual and sometimes revolutionary role reversal in which political processes; class alliances; the categories of rent, interest, merchants' profit, and taxation; and the assets of physical and social infrastructures were converted from interdependent though interlinked preconditions and determinants of political-economic processes into pure servants of capital accumulation. The role of the urban process, as well as the mechanisms of its development, shifted dramatically with this role reversal.

Primitive accumulation and other processes of appropriation do not guarantee, however, that the surpluses can be assembled in time and space in exactly the right proportions for strong capital accumulation to proceed. In eighteenth-century Britain, for example, the strong capital surpluses more than matched the surpluses of labor power. Wages rose, and much of the surplus was absorbed in consumption projects. In contrast, much of contemporary Africa, Asia, and Latin America is faced with a situation in which immense quantities of labor power have to be dispossessed to release very little capital, creating massive and chronic surpluses of labor power in a context of serious capital shortage. Situations can arise and even persist, therefore, in which surpluses of one sort cannot be absorbed because surpluses of another sort are not present in the requisite quantities and qualities. Under contemporary conditions, this means that either capital or labor power is devalued, but not both. To the degree that the dominant power relations favor capital and that the qualities of the latter can quickly adapt to shortages of labor power through technological innovation, so the likely persistent (as opposed to occasional) condition will be that of capital shortage and labor power surpluses. This is, for example, the hallmark of much of contemporary Third World urbanization.

The urban assembly of such surpluses does not guarantee, however, that they will be used capitalistically. We here encounter a historical problem of considerable political, social, and economic complexity. The most successful of the urban centers from the standpoint of assembling the surpluses often used their political power in ways inimical to the direct flowering of capital accumulation through production. The latter, after all, meant a major transformation of class power and structure and cutting loose any controls over technological innovations that might threaten, as they nearly always do, the value of any existing base of assets. In addition, the purpose of appropriation of surpluses was the building of wealth as a basis for conspicuous consumption, and it was not immediately apparent to those who held that wealth that the best way to preserve it was to use it as capital. The more powerful urban-based class alliances often used their class and monopolistic power to organize against the capitalism they helped spawn.

Unfortunately for the city states, the very methods employed in the assembly of much surpluses tended in the end to undermine their powers of monopolistic control over money, space, and commodity flow. Trade and commerce meant monetization, and that always has a dissolving effect upon the coherence of community (*Grundrisse,* 224–25; *Consciousness and the Urban Experience,* chap. 1). Anyone who holds money is perpetually tempted to use it for personal gain outside of the controlling powers of some urban-based class alliance (see Chap. 6). Trade also entailed the formation of other trading centers that could ultimately become rivals. And to the degree that new products and military technology were important facets of a city's power, so innovation became a vital force that no urban-based class alliance could afford to stifle if it was to prosper and survive. Competition between urban centers became a major check on internal monopoly controls and tended to create conditions of instability out of which the circulation of capital through production could more easily gain a foothold.

The separation of laborers from control over their means of production (through physical or legalized violence in some instances and out of attraction to the opportunities of urban life in others) and their conversion into wage laborers forms the other half of the conditions necessary to the rise of capital as a hegemonic mode of production and circulation. But again, there was no guarantee that displaced laborers would become wage laborers. The processes of displacement and socialization into the proletariat were anything but idyllic (Pollard 1965). The habituation of the worker to wage labor, the inculcation of a work discipline and all that went with it, and the formation of freely functioning labor markets were not, and still are not, easy matters. The workers themselves often sought and acquired corporatist powers that checked the liberty of labor and in so doing learned to support mercantilist and monopolistic practices on the part of their rulers. The urban-based guilds and the corporatist organization of labor also formed a powerful barrier to free capital accumulation (this remained a serious problem in France, for example, throughout the whole of the nineteenth century as the case of Second-Empire Paris clearly demonstrates – see *Consciousness and the Urban Experience,* chap. 3). The labor process tended to stagnate into monopolistic mosaics of craft control at the same time as labor markets froze into rigid configurations.

But again, there were processes at work that undermined guild, artisan, and craft controls. Immigration of displaced rural labor into urban centers meant that the pressure of labor surpluses was never absent. Competition between urban centers for new products and new technologies meant pressure (sometimes organized by the ruling class and therefore a major point of class struggle) to open up the labor process to new possibilities. And the wage laborers themselves, particularly if they aspired, as many did, to become small masters and entrepreneurs, could often undermine the corporatist logic. The original qualities of labor power and labor's powers of organization in the different industries nevertheless had deep impacts upon the prospects for using surplus wealth as capital in production. Hardly surprisingly, the pace of accumulation and technological innovation varied greatly from one urban center to another. Without the force of interurban competition, the pace of capitalist penetration into production would certainly have been much slower and may even have been blocked altogether.

For these sorts of reasons it proved easier for capitalist industrialization to emerge in entirely new urban centers in which the politics of monopoly control and the tactics of mercantilism were less firmly entrenched. In some instances the capitalist penetration of agriculture, coupled with technological innovation in the countryside, proved the cutting edge of capitalist development. The circumstances that gave free rein to the circulation and accumulation of capital through production, that allowed labor markets to function freely and new technologies and forms of labor organization to be deployed without restraint, were evidently diverse even if rather restricted (Merrington 1975; Holton 1984). The urban centers had, nevertheless, crucial advantages in relation to accumulation. The vast assembly of assets in the built environment, though oriented primarily to trade, consumption, and political-military dominance, could be converted almost costlessly into assets for capital accumulation – the consumption fund could be transformed into fixed capital in the built environment simply by changing patterns of use (see Chap. 1). The transport and communications systems built to facilitate appropriation, trade, consumption, and military control could likewise be used by capitalist producers. Countries like Britain and France in the eighteenth century that had vast assets of this sort were, therefore, in a far better situation for capitalist development than many contemporary Third World countries whose asset base is extremely limited. Furthermore, the sociopolitical institutions, private property rights, and systems of command of money (banking and treasuries) could also be mobilized behind the geopolitics of capital accumulation as command centers for the circulation of capital. The breakthrough into a predominantly capitalist mode of production and circulation was not, therefore, a purely urban or a purely rural event. But without the urban accumulation of surpluses of both capital and labor power, one of the crucial necessary conditions for the rise of capitalism would not have been fulfilled.

The transition to a capitalist mode of production was signaled by a shift from the production of capital and labor surpluses by processes external to the circulation of capital to an internalization of surplus production within the circulation of capital itself. That shift was also signaled, I have argued, by a role reversal in which rent, interest, merchants' profit, state powers and functions, and the production of built environments became servants of capital accumulation and subservient to its dominant logic.

Consider, for example, the manner of production and use of the physical and social landscape necessary to capital accumulation. The story of how that landscape is produced and used is central to my theorization of the urban process. I focus upon it because it is the product of a process – or a set of processes, for we here confront matters of great intricacy and complexity – that gives definite shape and form to a capitalist urbanization process that would otherwise appear far more flexible and fluid than is in fact the case. The transition from a historical condition in which that landscape is produced by forces outside the logic of accumulation to one in which it is integrated within that logic is signaled when the circulation of capital produces the necessary surpluses and the sufficient conditions for the shaping of physical and social space within its confines. And that occurs when overaccumulation begins to grab hold of immediate production and consumption – when, in short, crises become clear manifestations of the internal contradictions of capitalism rather than being reasonably attributable to external circumstances of natural calamity (such as harvest failures) or social breakdown (wars of aggrandizement or internal civil and political strife). Though hints of that occurred before, 1848 was perhaps the first indisputable and unambiguous manifestation of that kind of crisis within the capitalist world.

The production of the physical and social landscape of capitalism was thereafter increasingly caught up in the search for solutions to the overaccumulation problem through the absorption of capital and labor surpluses by some mix of temporal and geographical displacement of surplus capital into the production of physical and social infrastructures (the "secondary" and "tertiary" circuits depicted in fig. 3). I have dwelt at length (Chaps. 1 and 2; see also Harvey 1982, 1985) on the potentialities and limitations of that process. Suffice it to remark that the problems of overaccumulation and devaluation are thereby imparted to the physical and social landscape so that its whole historical evolution dances to their tune. For that to happen, however, a whole set of preconditions has to be realized, including, of course, that most essential precondition of all, the command of immediate production and consumption by the industrial capitalist. It is, then, to that issue and its requisite form of urbanization that we now turn.

III. THE CAPITALIST PRODUCTION OF SURPLUSES AND THE INDUSTRIAL FORM OF URBANIZATION

The rise of the industrial city signaled the penetration of capital circulation into the heart of immediate production and consumption. The shift from the appropriation of surpluses through trade, monopoly, and military control to the production of surpluses through command over labor processes in production was slowly wrought. Not all sectors were immediately captured, of course (agriculture remained notoriously recalcitrant, finally succumbing in the advanced capitalist countries only after World War II and then often unevenly). But for those sectors subsumed, there was a dramatic transformation in the organization of the sociotechnical conditions of production and in the functioning of labor and commodity markets.

This meant that the whole basis of urbanization had to change. The preindustrial city had to be disciplined, weaned away as it were from its mercantilist proclivities, its monopolistic practices, and its assertion of the primacy of place over a capitalist organization of space in which relative rather than absolute locations had to dominate. The incorporation of the city state within the broader configurations of nation states – a tension that Braudel (1984) makes much of – was one important step in the direction that allowed the freer functioning of labor, commodity, and credit markets as well as the freer flow of capital and labor power between sectors and regions. Industrial capitalists, seeking out new resource bases and new sociotechnical conditions of production within entirely new urban areas outside the monopoly controls so prevalent in preexisting urban centers, could do so only in a context where a relatively strong nation state had secured the political and institutional basis for private property and that sort of control over the means of production which allowed the exploitation of labor power. Where industrial capitalism was grafted on to older structures (as in Paris and London), it assumed qualities quite different from those of the burgeoning capitalist industrialism of a Manchester or a Birmingham. There are, indeed, those who attribute the relative stagnation of capitalism in France to the inability to break with preexisting patterns of urbanization and the political power of prevailing urban-based class alliances (St. Etienne was the only new major city opened up in the nineteenth century). The story in Britain, Germany, and the United States was very different – new industrial centers opened up all over the place under the watchful institutional and legal eye of strong state power.

The industrial city was a new centerpiece of accumulation. The production of surpluses through the direct exploitation of living labor in production was its trademark. This meant the geographical concentration of labor power and productive force (epitomized in the factory system) and open access to the world market, which, in turn, meant the consolidation of universal money and credit. It meant, in short, the firm implantation of all those features of geographical and temporal organization of the circulation of capital that I began by describing. The geographical patterning of labor and commodity markets, of spatial and social divisions of production and consumption, and of differentiated sociotechnical mixes within the labor process became much more pronounced within the urban landscape. Intercapitalist competition and class struggle pushed the whole dynamic of urbanization toward the production of rational physical and social landscapes for capital accumulation. The search for profitable trade-offs between command over and creation of advantageous locations, coupled with adaptations in the sociotechnical conditions of production, became a much more visible moving force within the urban process.

The capacity of any urban-based class alliance to wield monopoly control, either internally or on the world stage, diminished. This is not to say that certain of the more important industrial centers – like Manchester in the nineteenth century and Detroit in the twentieth – did not assemble sufficient power to mimic for a while the behavior of urban-based class alliances in preceding eras. But such delusions of geopolitical grandeur soon foundered on the dynamics of growth and geographical expansion, technological change and product innovation, class struggle, international competition within a shifting frame of relative space shaped by revolutions in transport and communications, and the growing disruptions of crises of overaccumulation. Although the industrial city was a centerpiece of accumulation and surplus production, it has to be seen as a distinctive place within the spaces of the international division of labor, a mere element within a more and more generalized capitalist system of uneven geographical development.

But urban organization was vital to the working out of such a process. Though individual cities (like individual firms) exercised less and less control over aggregate processes and outcomes, their individual performance in the context of interurban competition set the tone, pace, and direction of historical-geographical development. The industrial city became, in short, a concrete means toward the definition of abstract labor on the world market (see Harvey 1982, 422–26). Value on the world market then became the standard against which every industrial center's performance was judged. The conception of the industrial city as a competitive unit within the uneven geographical development of global capitalism made more and more sense under such conditions.

Denied the grandeur of geopolitical posturing (a role increasingly reserved for nation states or capital cities like Paris), the tasks of urban politics within the industrial city had to shift toward more mundane concerns. The problems of organization and control, of management of physical and social infrastructures, were radically different from anything that had gone before, while the context of class alliance formation changed because class structures were redefined. Class warfare between capital and labor and the drive to reproduce that basic class relation of domination became the pivot of urban politics. The formation of physical and social infrastructures adequate to support the reproduction of both capital and labor power while serving as efficient frameworks for the organization of production, consumption, and exchange surged to the forefront of political and managerial concerns. Such problems had to be approached with an eye to efficiency and economy because that was the way to assure growth, accumulation, innovation, and efficiency in interurban competition. Public investments also had to be organized on an increasing scale and on a more and more long-term basis and in such a way as to compensate for individual capitalists' underproducing collective infrastructures.

It was precisely around such themes that Joseph Chamberlain built such a powerful class alliance in Birmingham in the 1860s, comprising representatives of industry, commerce, and the professions, with a great deal of solid working-class support. The emergence of a distinctive civic tradition in Leeds, Manchester, and Birmingham during the nineteenth century – a process ultimately paralleled in many a new industrial city – was part of a search to define a new urban politics appropriate to new circumstances. Older cities, like Haussman's Paris (see *Consciousness and the Urban Experience,* chap. 3), had to acquire the same virtues of efficiently organized capitalist modernity by a far more tortuous route. The bases for class-alliance formation and confrontation were very different and the objectives equally so. But the common problems faced (from debt-financing infrastructural investments to finding ways to rationalize urban space as a whole) and the common techniques employed (engineering skills merging into rational urban planning) also induced a certain tactical convergence toward a distinctively capitalist kind of urban managerialism.

Industrial capitalism also wrought far-reaching transformations of all aspects of civil society. Traditional social relations of work were altered or destroyed and new social structures forged against the background of freely functioning labor markets and powerful currents of technological change. Integrating immigrants and absorbing the shocks of technological change posed key questions of socioeconomic policy and political management. The role of women changed in both labor markets as well as in the household, and the family had to adapt and reconstitute itself to the buying and selling of labor power as a way of life. At the same time, social reproduction processes had to incorporate mechanisms directed toward the production of labor supply with the right qualities and in the right quantities. Attention had to be paid to such questions at a time when the bonds of civil society threatened to break asunder under the strain of the alienations of class conflict, the anomie of individualistic labor markets, and the sheer rage at the new regimes of domination. It took real political talent and much subtle maneuvering to keep the urban pot from boiling over under the best of conditions, and the new bourgeoisie had to find new ways to keep the revolutionary turmoil under control. Bourgeois surveillance of the family and interventions in the cultural, political, and social milieu of the working classes began in earnest. Above all, the ruling-class alliance had to find ways to invent a new tradition of community that could counter or absorb the antagonisms of class. This it did in part by accepting responsibility for various facets of social reproduction of the working class (health, education, welfare, and even housing provision) and mobilizing sometimes brutal and sometimes subtle means of social cooptation and control – police, limited democratization, control of ideology via the churches or through the newly emerging organs of mass communication, and the manipulation of space as a form of social power. And the working class, as part of its own tactic of survival, also sought a new definition of community for itself. With its help, industrial capitalism in fact forged, with amazing rapidity, new traditions of urban community out of conditions of social disintegration and class conflict.

So strong did the popular attachments become that they formed a major barrier to further urban transformations within the short space of a generation.

But that sense of urban community, along with the "structured coherence" (see Chap. 6) of the sociotechnical conditions of production and consumption and of labor supply in relation to industial capitalism's needs, could never become a stable configuration. The dynamics of accumulation and over-accumulation, of technological change and product innovation, of shifting international competition in a rapidly changing structure of relative space (transformed through revolutions in transport and communications), kept even the best-managed and the most efficiently organized industrial city under a perpetual cloud of economic uncertainty. The mobilities of capital and labor power could not be controlled – this was, after all, the essence of free-market capitalism – nor could the context of wage or profit opportunities elsewhere. Each and every participant in or supporter of some urban-based class alliance faced the temptation of abandoning or undermining it for individual gain.

It was within this space of relative uncertainty and insecurity that a relatively autonomous urban politics made its mark. A charismatic leadership (sometimes collective and sometimes individual) could build its reputation on successful strategies for progress and survival in an uncertain and highly competitive world. Strategies could vary from ruthless creative destruction of anything that stood in the way of capitalist rationality, modernity, and progress to attempts to insulate against or even break out from under the coercive laws of competition through movements toward municipal socialism. But the latter could always be checked by two reserve powers. The discipline of competition and of "abstract labor" on the world market could not for long be held at bay without lapsing into an isolationism that could destroy much that had been achieved. Political experience taught the bourgeoisie another lesson that could be used to check the undue radicalism of any urban-based political movement: superior control over space provided a powerful weapon in class struggle. The Parisian revolutions of 1848 and 1871 were put down by a bourgeoisie that could mobilize its forces across space. Control over the telegraph and flows of information proved crucial in disrupting the rapidly spreading strike of 1877 in the industrial centers of the United States. Those who built a sense of community across space found themselves with a distinct advantage over those who mobilized the principle of community in place. Politically, this meant increasing ruling-class reliance upon national and, ultimately, international power sources and the gradual reduction of the sphere of relative autonomy of urban-based class alliances. It was no accident, therefore, that the nation state took on new roles and powers during that period of the late nineteenth century when diverse movements toward municipal socialism and machine politics with a working-class or immigrant orientation gathered momentum. The more the bourgeoisie lost control over urban centers, the more it asserted the dominant role of the nation state. It reinforced the authority of the spaces it could control over the places it could not. This was the political lesson that the bourgeoisie learned from the rise of the industrial city as a powerhouse of accumulation and a crucible of class struggle.

The industrial city was, therefore, an unstable configuration, both economically and politically, by virtue of the contradictory forces that produced it. On the one hand it sought and sometimes achieved a rational internal ordering to facilitate space-time coordinations in production, in flows of goods and people, and in necessary consumption coupled with the ordered construction of social spaces for the reproduction of labor power and well-managed patterns of social provision, built-environment production, and urban political management. From this standpoint it appeared as a relatively efficient corporation geared for competition on the stage of world capitalism. On the other hand, the industrial city was beset with the social anarchy generated by crises of overaccumulation, technological change, unemployment and de-skilling, immigration, and all manner of factional rivalries and divisions both within and between social classes. Interurban competition to some degree exacerbated the difficulties, since it increased the pressures toward product innovation and technological change. The industrial city had to consolidate its function as an innovation center if it was to survive. But innovation brought disruption and also lay at the root of the overaccumulation problem. The industrial city, as a powerhouse of accumulation and innovation, had to be the prime vehicle for the production of overaccumulation.

How were the prodigious surpluses of capital, and to a lesser extent of labor power, to be absorbed without devaluation and destruction? The periodic crises of industrial capitalism indicated no easy answer to that question. Surpluses could be and were in part absorbed by deepening productive forces (including those of labor power) within the industrial city through an increasing flow of investments into long-term physical and social infrastructures (see Chap. 1). They were also absorbed through geographical expansion (see Chap. 2). The search for a "spatial fix" for the overaccumulation problem spawned industrial development in far-off lands and the increasing linkage of urban industrialism into a system of urban places through movements of money, capital, commodities, productive capacity, and labor power. That way the threat of overaccumulation could be staved off, sometimes at the cost of primitive accumulation from precapitalist societies or forcible implantation of capitalist industrialization on societies (like the United States) that had sought a radically different path to social progress. The industrial city had to be, therefore, an imperialist city. And if it wanted to retain its hegemonic competitive position within a proliferating world market, it had to be prepared to conjoin political and military imperialism with an economic imperialism that rested on technological superiority and innovation coupled with superior organization of production, capital markets, and trade within the social and geographical divisions of labor. Joseph Chamberlain even made such themes central to the ideology of the class alliance (including many workers) that he kept together in Birmingham in the troubled depression years of the 1880s and 1890s.

But interurban competition, spiraling technological innovation and overaccumulation, and geographical expansionism constituted an unstable mix. Indeed, this was the kind of underlying pressure that produced national geopolitical rivalries and two world wars, the second of which inflicted enormous and uneven geographical destruction on urban assets – a neat but hideously violent resolution to capitalism's overaccumulation problem. Was there any way to avoid such a paralyzingly destructive resolution of capitalism's internal contradictions?

IV. THE ABSORPTION OF SURPLUSES: FROM FORDISM TO THE KEYNESIAN CITY

Underconsumption seemed to be, and in a sense was, the reverse side of the coin to overaccumulation. If that was so, then why could not the contradictions of capitalism be resolved by closer attention to the expansion of consumption, particularly on the part of the working masses of the population who were, in any case, not only economically needy but politically aggressive? The search for a solution of that sort underlay a shift in focus from production to distribution and consumption. Capitalism shifted gears, as it were, from a "supply-side" to a "demand-side" urbanization. Let us consider the elements of that transition.

The rise of the corporation from the ashes of the family firm, coupled with major reorganizations of labor processes in many industries, liberated many aspects of production from reliance upon access to particular natural or urban assets. Industry became increasingly footloose, not above the calculus of local advantages of labor supply or social and physical infrastructures, but more able to exploit their uneven availability within the urban system. This did not automatically produce geographical decentralization of production under unified corporate control. Precisely because much of the impetus toward the formation of large corporations, trusts, and cartels came from the need to curb excessive competition, the emphasis lay on the joys of monopoly rather than the rigors of competition. And monopoly powers could be used geopolitically, either to further concentrate production geographically or to protect geographical concentrations already achieved. The distortion of relative space imposed by the United States steel companies through their "Pittsburgh plus" system of steel pricing was one example of many sustained attempts to use monopoly power to protect a particular urban region against external competition. It took many years, and in some cases deep financial trauma, for large corporations to learn how to internalize competition (between, for example, regional branch plants) and use their power to command space and manipulate geographical dispersal to their own advantage. In this they were always limited, of course, by the need to assure internal economies of scale and continuous flow in production while sustaining reasonable proximity to networks of subcontractors and adequate labor supplies.

Relieved of the burden of excessive competition in production, the large corporations became much more sensitive to the control of labor power and markets as the basis for a constant and secure flow of revenues and profits (Gramsci 1971). Their attachment to large-scale production also led them to direct their attention to mass rather than privileged and custom markets. And the mass market lay within the working class. This was the basis for Fordism. Increased productivity in the workplace was compensated by higher wages that would allow the workers to buy back a larger share of the commodities they helped produce. Ford himself was quite explicit about that strategy when he inaugurated the eight-hour, five-dollar work day at his auto plant in 1914. But to the degree that workers are never in a position to buy back the whole of the output they create, so the large corporations were forced into strategies of geographical dispersal in order to ensure market control on an expanding basis. And it was not long before the advantages of decentralization of component production as well as of final assembly became apparent. But these adjustments were slowly wrought, depending to a considerable degree upon the changing space relations created by new systems of transport and communications.

The more corporations used their powers of dispersal, however, the less urban regions competed with each other on the basis of their industrial mix and the more they were forced to compete in terms of the attractions they had to offer to corporate investment as labor and commodity markets and as bundles of physical and social assets that corporations could exploit to their own advantage. The corporations became less and less place-bound and more and more representative of the universality of abstract labor on the world market. Innovation likewise tended to shift its breeding ground from the interstices of the urban matrix into government and corporate research labs, though new product innovation still retained some of its more traditional urban bases.

The growing power of the credit system added its weight to these shifts. The centralization of credit power was nothing new – the Barings and the Rothschilds early learned that superior information and capacity to deploy money power over space gave them the power to discipline even nation states throughout much of the nineteenth century. But they had largely confined their operations to government debt and selected large-scale projects, like railroads, leaving commercial and industrial credit and consumer loans (if they existed at all) to other, more fragmentary sources. The manifestation of crises in the nineteenth century as credit and commercial crises – 1847–48 being a particularly spectacular example – prompted major changes in capital and credit markets. The stock market and the reorganization of banking changed the whole context of credit and finance by the end of the nineteenth century. The rise of finance capital (see Harvey 1982, chap. 10) had all manner of implications. It facilitated the easier movement of money capital from one sector of production or geographical region to another and so allowed the much finer tuning of the relations between the social and geographical divisions of labor. It made the debt-financed production of urban infrastructures that much easier, as well as facilitating the production of long-term investments that reduced spatial barriers and helped further annihilate space with time. It therefore meant a smoother and accelerating flow of capital into the deepening and geographical widening of urban infrastructures at the very moment when increasingly footloose corporations were looking to tap into the particular advantages to be derived from those sorts of investments. The effect, however, was to tie the production of urban infrastructures more tightly into the overall logic of capital flow, primarily through movements in the demand and supply of money capital as reflected in the interest rate. The "urban construction cycle" therefore became much more emphatic, as did the rhythmic movement of uneven urban development in geographic space (see Chap. 1).

But the credit system also seemed to pack another punch, one that could virtually annihilate the overaccumulation problem at one blow. The proper allocation of credit to production and consumption held out the prospect of balancing both within the constraints of continuous profit realization. The flow of money and credit into production had simply to be matched by the flow of money and credit to support consumption in order that self-sustained growth continue in perpetuity. There were, of course, many problems to be resolved. Balanced growth could not be achieved through any pattern of production and consumption if accumulation was to be achieved and profits realized. The proper balance between productive consumption (investments that enhanced the capacity of the productive forces) and final consumption (investments and flows that enhanced the living standards of the bourgeoisie as well as of the working class) had to be struck. But the credit system nevertheless seemed to have the potential power to do what individual corporations seeking a Fordist compromise set out to do but could not because of their limited power to affect distribution. To the degree that the credit system became oriented to these tasks it became the major vehicle for the transformation to demand-side as opposed to supply-side urbanization.

But there were two interrelated problems. First, financial markets, like money itself, embody immense powers of centralization in the midst of the greatest possible dispersal of powers of appropriation. This permits the concentration of key decision-making functions for global capitalism in a few hands (like J. P. Morgan) in a few urban centers (like New York and London). This poses the threat of private perversion of this immense centralized social power for personal gain or the use of monopoly power for narrow geopolitical ends. It also tends to consolidate the hierarchical geographical ordering of financial centers into a system of authority and control that is as much self-serving as it is facilitative of balanced accumulation. Worse still, and this brings us to the second objection, the formation of "fictitious capital" (all forms of debt) has somehow to be regulated if it is not to spiral out of control into orgies of speculation and unchecked debt creation (see Harvey 1982, chaps. 9 and 10). How, for example, was the debt on urban infrastructures to be paid off if the latter did not enhance surplus value production? And if such investments were productive, would not that merely exacerbate the overaccumulation problem? Periodic financial crises indicated that overaccumulation could all too easily be translated into an overaccumulation of debt claims on nonearning assets.

It is against this background that we have to understand the increasing pressure toward state intervention in macro-economic policy. It was, of course, to the nation state that the bourgeoisie turned, in part because this was the space they could most easily control but also because the nation state was the institutional frame within which fiscal and monetary politics were traditionally formulated. It was the switch into Keynesian strategies of fiscal and monetary management that consolidated the turn to demand-side urbanization. The trauma of 1929–45 provided the catalyst. When the depression hit in the United States, Ford, true to his colors, saw it as an underconsumption problem and tried to raise wages. Forced within six months by the logic of the market to back down, Fordism failed and had to convert itself (often reluctantly) into state-managed Keynesianism and New Deal institutional reforms and politics. For more than a generation, capitalist urbanization (particularly in the United States) was shaped after the added trauma of World War II into a state-organized response to what were interpreted as the chronic underconsumption problems of the 1930s.

The implications for the urbanization of capital were profound. The Keynesian city was shaped as a consumption artifact and its social, economic, and political life organized around the theme of state-backed, debt-financed consumption. The focus of urban politics shifted away from alliances of classes confronting class issues toward more diffuse coalitions of interests around themes of consumption, distribution, and the production and control of space. The "urban crisis" of the 1960s bore all the marks of that transition. The shift also provoked a serious tension between cities as "workshops" for the production of surplus value and cities as centers of consumption and realization of that surplus value. There was a tension between the circulation of capital and the circulation of revenues, between the spatial division of labor and the spatial division of consumption, between cities and suburbs, and so forth. Keynesian policies radically changed, in fact, the manner and style of temporal (debt-financed) and spatial displacement of the overaccumulation problem. Let us see how.

Unlimited temporal displacement could be achieved to the degree that state-backed credit allowed the unlimited creation of fictitious capital. Keynes had meant deficit financing as a short-run managerial device, but permanent and growing deficits were built up as the business cycle was kept under control and the urban construction cycle that had been so powerfully present before 1939 was all but eliminated. Overaccumulated capital and labor power were switched into the production of physical and social infrastructures; and if such investments helped produce more surpluses, then another round of switching could take place. The prospect arose, for urban regions as well as for nations, of a permanent upward spiral of economic growth, provided, of course, the targets of debt-financing were well chosen. Investments in transportation, education, housing, and health care appeared particularly appropriate from the standpoint of improving labor qualities, buying labor peace, and accelerating the turnover time of capital in both production and consumption. But the process rested on unlimited debt creation no matter how it was worked out. By the 1970s, the United States was weighed down by what even *Business Week* conceded was a "mountain" of public, private, and corporate debt, much of it wrapped up in urban infrastructures. The accumulation of debt claims posed a problem. The attempt to monetize them away produced strong surges of inflation, thus demonstrating that the threat of devaluation of commodities and other assets could be converted into the devaluation of money (cf. Harvey 1982, chap. 10). But any counterattack against inflation could only put a great deal of urbanized capital at risk. The collapse of the property market worldwide in 1973 (and the collapse of banking and financial institutions heavily caught up in property finance) and the New York fiscal crisis of 1974–75 were opening gambits in a whole new mode of the urban process based on non-Keynesian approaches.

The temporal displacement of overaccumulation through debt-financed infrastructure formation was accompanied by strong processes of spatial reorganization of the urban system. Long reduced to a commodity, a pure form of fictitious capital (see Chap. 4), land speculation had also been a potent force making for urban sprawl and rapid transitions in spatial organization, particularly in the United States. The means of further dispersal – the automobile – had also been on hand since the 1920s. But it took the rising economic power of individuals to appropriate space for their own exclusive purposes through debt-financed homeownership and debt-financed access to transport services (auto purchases as well as highways), to create the "suburban solution" to the underconsumption problem (Walker 1976, 1981). Though suburbanization had a long history, it marked post-war urbanization to an extraordinary degree. It meant the mobilization of effective demand through the total restructuring of space so as to make the consumption of the products of the auto, oil, rubber, and construction industries a necessity rather than a luxury. For nearly a generation after 1945, suburbanization was part of a package of moves (the global expansion of world trade, the reconstruction of the war-torn urban systems of Western Europe and Japan, and a more or less permanent arms race being the other key elements) to insulate capitalism against the threat of crises of underconsumption. It is now hard to imagine that postwar capitalism could have survived, or to imagine what it would have now been like, without suburbanization and proliferating urban development.

The whole process rested, however, on continuous and radical restructurings of the space-time matrices that frame economic decisions as well as social and political life. The revolution in space relations overwhelmed the punctiform settlement patterns of industrial capitalism and replaced them with "space-covering" and "space-packing" patterns of labor and commodity markets merging into pure megalopolitan sprawl. The urban-rural distinction was swamped with respect to production in the advanced capitalist societies, only to be reproduced as an important consumption option. Geographical dispersal and space-packing had its limits, however. The more investments crystallized into fixed spatial configurations, the less likely it became that space could be further modified without being devalued. This was not a new problem. The reshaping of the industrial city to fit Keynesian requirements imposed economic costs and sparked social resistance, often on the part of working-class communities that had forged their identity from the industrial experience. Greater attachment to that sense of community (and a reluctance to treat land as pure fictitious capital) slowed the pace of suburbanization in Europe, perhaps slowing overall growth as a result. But even in the United States, the erosion and occasional destruction of the preexisting bases of community in older areas became widely seen as the negative side of the golden currency of suburbanization. As the processes of spatial transformation gathered pace, so those problems became more widespread, affecting middle and even upper-income communities with much more power to resist.

The Keynesian city put much greater emphasis upon the spatial division of consumption relative to the spatial division of labor. Demand-side urbanization depended on the mass mobilization of the spirit of consumer sovereignty. Surpluses were, in effect, widely though unevenly distributed, and the choice of how to spend them was increasingly left to the individual. The sovereignty, though fetishistic (in Marx's sense), was not illusory and had important implications (see *Consciousness and the Urban Experience,* chap. 5). Since there are no natural breaks on the continuum of money power, all kinds of artificial distinctions could be introduced. New kinds of communities could be constructed, packaged, and sold in a society where who you were depended less and less on class position and more and more on how you spent money in the market. Living spaces were made to represent status, position, and prestige. Social competition with respect to life-style and command over social space and its significations became an important aspect of access to life chances. Fierce struggles over distribution, consumption rights, and control over social space ensued. Once largely confined to the upper layers of the bourgeoisie, such struggles now became part of urban life for the mass of the population. It was largely through such struggles and the competition they engendered that demand-led urbanization was organized to capitalistic ends.

Urban politics had to change its spots. The success of the Keynesian project depended upon the creation of a powerful alliance of class forces comprising government, corporate capital, financial interests, and all those interested in land development. Such an alliance had to find ways to direct and channel a broadening base of consumer sovereignty and increasing social competition over consumption and redistribution. It had to shape and respond to the quest for new life-styles and access to life chances so as to create patterns of temporal and spatial growth conducive to sustained and reasonably stable capital accumulation. But the basis of popular legitimacy (at both the local and national levels) had to rest on performance with respect to distribution and satisfaction of consumer wants and needs. While there were phases of concordance of these two aims, there were also serious points of tension.

The attempt to use the urban process as a vehicle of redistribution ran up against the realities of class structure, income differentials, and minority deprivation. The strong processes of spatial reorganization of consumer landscapes left behind growing pockets of abandonment and deprivation, for the most part concentrated in inner cities. It was almost as if creative destruction split into the physical and social destruction of the inner cities and the creation of the suburbs. But all was not necessarily well at the other end of the social scale. As consumers, even upper echelons of the bourgeoisie could demand protection against developers and others who wanted to shape space for growth and profit. Peculiar kinds of "consumer socialism," built around the power of the local government to check growth-machine politics, could take root even in affluent areas (such as Santa Monica). Consumer sovereignty, if taken seriously, presupposes a certain popular empowerment to shape the qualities of urban life and construct collective spaces in an image of community quite different from that embodied in the circulation of capital. The production of space tended to run up against sensitivity to place. The boundary between consumerist innovation promoted by capitalism and attempts to construct community in the image of real self-fulfillment became exceedingly fuzzy.

It was in exactly such a context that the inner-city uprisings of the 1960s (and some of the later urban unrest in Europe) coupled with no-growth and environmentalist movements on the fringes checked the accelerating trajectory of urban transformation typical of the Keynesian city. The urban social movements of the 1960s focused strongly on distribution and consumption issues, and urban politics had to adjust from a pure growth machine track to redistributive issues. The circulation of revenues had to be managed so as to ensure economic and political inclusion of a spatially isolated under-class and a socially just distribution of benefits within the urban system. The city was increasingly interpreted as a redistributive system. Questions of jobs and employment and of the city as an environment for production, though never excluded from consideration, were viewed as minor elements in a complex matrix of forces at work within the urban process. Rivalry over the circulation of revenues and redistributions tended, however, to exacerbate intercommunity tensions and geopolitical conflicts (between, for example, cities and suburbs). And there was nothing about such a strategy that necessarily assured smooth sailing for the circulation of capital either.

Three problems were central to the temporal and spatial displacement of overaccumulation through demand-side urbanization. First, temporal displacement led to increased indebtedness and strong inflationary pressures. A return to classic forms of devaluation would, however, have put vast urban investments at risk and would have destroyed well-established patterns of redistribution, thus making such a policy reversal harder and harder to confront as time went by. Second, investment in suburban sprawl and the "space-covering" style of urbanization entailed the fixation of fragmented spaces within which the drive toward local empowerment and community formation created barriers to the further pursuit of the suburban solution and the spatial fix. The process of spatial displacement either slowed or was forced to ever more intense levels of creative destruction and contentious devaluation. Third, demand-led urbanization (with all of its concerns for individualism, consumer sovereignty, life-style and status, and social competition for command over space) pushed the focus of concern away from the direct circulation of capital toward the circulation of revenues. It emphasized the production of preconditions for the spatial division of consumption rather than of production. This shift was as dangerous as it was provocative. It assumed an automatic and apropriate supply-side response to match the debt-financed growth of effective demand. The tension between cities as "workshops" for production and as centers for consumption was not easy to contain. Investment in the physical and social infrastructures for consumption, coupled with the politics of redistribution, does not necessarily create a favorable climate for capitalist production. And since corporations now possessed increased powers of geographical mobility, and since finance capital had by now become extraordinarily mobile, cities became much more vulnerable to job loss, capital flight, and corporate disinvestment. This was to be the dilemma of the 1970s, though evidence of it could be seen much earlier.

This account of demand-side urbanization and its inner tensions is, admittedly, highly simplified and rather biased toward the American case. It is also rather superficial to the degree that it does not pay sufficient attention to the necessary unity of production and consumption within the logic of the production and realization of surplus value. That question was never far from the surface of concerns in the midst of industrial urbanization. Engels had certainly noted it in his examination of Manchester in 1844, in his celebrated description of the different residential zones of consumption that reflected class relations in production. Urban proletariats had long formed significant captive markets to which capitalists catered, and the question of the importance of local effective demand as the basis of a vigorous export trade had long been broached. And then there were those cities, like Paris or London, that traditionally functioned as centers of conspicuous consumption, and where the volume and type of effective demand were critical in setting the tone and pace of local industrial activity.

The Keynesian city was not blind to questions of production either. But there was a shift of emphasis which was of sufficient proportions to warrant depiction as a major transformation of the urban process. Though the Great Depression was much more than a crisis of underconsumption, the fact that it appeared as such and that the capitalist class responded to it as such laid the groundwork for a totally new patterning of the urban process. Nor does it matter that urbanization as a whole cannot survive without some consideration of cities as workshops for production if the whole response to underconsumption problems is to strive to create a "post-industrial city" in which industrial development has no role. The production of the Keynesian city was a real response to the surface appearance of underconsumption as the root of capitalism's problems. That real response to a surface appearance created, of course, as many problems as it solved.

Demand-side urbanization produced a very different-looking city of low-density sprawl, distinctive spaces of consumption (ranging from produced rural bliss to intense in-town living separated by what increasingly appeared as the no man's land of the suburb), and strange significations of life-style and social status etched into a landscape of unrelieved consumerism. Production increasingly meant the production of space and of long-term investments, behind which stood powerful growth coalitions that managed the new-style urbanization of capital in ways symbiotic with their own interests. They needed new instrumentalities in the realms of finance capital and the state and strong powers of persuasion and ideological control to ensure that consumer sovereignty was sovereign in the right way, that it produced "rational consumption" in relation to accumulation through the expansion of certain key industrial sectors (autos, household equipment, oil, and so forth). The Keynesian city increasingly appeared, then, as a post-industrial city, as a consumption artifact nourished by service provision, information processing, and the support of command functions in government and finance.

The politics also changed. The class relations of production were partially masked by artificial marks of consumption while struggles over distributive shares and the control of social space generated a troublesome factionalism that had the fortunate side-effect of permitting the ruling-class alliance to divide and rule with relative ease. The basis of political legitimacy shifted from managing class relations toward distributive justice and a not necessarily compatible concern to satisfy consumer desire and sovereignty. Fights over the control of social space (some progressive and other reactionary) and the increasing cleavage between city and suburb produced new lines of geo-political tension. The urban crises of the 1960s were built out of exactly such ingredients as these. There were fights over consumption (individual and collective) and distribution as well as struggles over command of social space and what it contained. And the whole style of thinking about urbanization followed suit (see Chap. 7). The literature of that era on the delivery of health care, education, transportation and welfare, and on the rational organization of space for accumulation as well as on the resolution of intercommunity conflict, reflected a style of urbanization in which questions of production and fundamental class relations were held in abeyance, a constant backdrop to a foreground of quite different political and economic concerns.

V. BALANCING SURPLUS PRODUCTION AND ABSORPTION: THE STRUGGLE FOR URBAN SURVIVAL IN THE POST-KEYNESIAN TRANSITION

The collapse of the Keynesian program changed all that. Each of the pinions of the postwar strategy for avoiding the dangers of underconsumption eroded during the late 1960s. The revival of world trade through international capital flow led to a proliferation of the overaccumulation problem. Competition from Western Europe and Japan sharpened as the capacity to absorb further investments profitably fell. Inflationary financing appeared to resolve the difficulty by provoking a wave of international lending that was to lie at the root of the subsequent monetary difficulties (the instability of the dollar as a reserve currency) and the international debt crisis of the 1980s. The same policies generated a spiraling flow of surplus capital and labor power mainly into the production of urban built environments (property investment, office construction, housing development) and to a lesser degree into expansions of the social wage (education and welfare). But when monetary policy was tightened in response to spiraling inflation in 1973, the boom of fictitious capital formation came to an abrupt end, the cost of borrowing rose, property markets collapsed (see fig. 9), and local governments found themselves on the brink of, or in New York's case plunged into, the traumas of fiscal crisis (no mean affair when we consider that New York City's budget and borrowing were far greater than those of most nation states). Capital flows into the creation of physical and social infrastructures (the secondary and tertiary circuits of fig. 3) slowed at the same time as recession and fiercer competition put the efficiency and productivity of such investments firmly on the agenda. That there had been and were serious problems of overaccumulation of assets in the built environment and of obligations in the field of social expenditures became apparent for all to see. Much of the investment was producing a very low rate of return, if any at all. The problem was to try to rescue or trim as much of that investment as possible without massive devaluations of physical assets and destruction of services offered. The pressure to rationalize the urban process and render it more efficient and cost-effective was immense (see Chap. 7).

The running out of steam of demand-side urbanization was powerfully intermingled, therefore, with the grumbling economic problems of the 1970s and 1980s. And to the degree that urbanization had become part of the problem, so it had to be part of the solution. The result was a fundamental transformation of the urban process after 1973. It was, of course, a shift in emphasis rather than a total revolution (in spite of what supply-siders and neoconservatives proclaimed on both sides of the Atlantic). It had to transform the urban legacy of preceding eras and was strictly limited by the quantities, qualities, and configurations of those raw materials. It occurred in fits and starts, dancing uncertainly to the seemingly arbitrary shifts in monetary and fiscal policy and the strong surges of international and interurban competition within the social and spatial divisions of labor. It also had to move tentatively in the face of uncertain powers of popular resistance. And it was not clear how, exactly, the urbanization of capital should adapt to problems that were anything but underconsumption problems. The problems of stagflation could be resolved only through a closer equilibrium between the production of surpluses and their real as opposed to fictitious absorption. The question of the proper organization of production came back center stage after a generation or more of building an urban process around the theme of demand-led growth. How could urban regions blessed largely with a demand-side heritage adapt to a supply-side world?

Four different possibilities, none of them mutually exclusive and none of them costless or free of serious political and economic pitfalls, seemed possible. I consider each in turn. For the sake of clarity, I shall consider them from the standpoint of urban regions as competitive economic and geopolitical units within a capitalist geography of seesawing uneven development (see Chap. 6; Smith 1984).

Competition within the Spatial Division of Labor

Urban regions can seek individually to improve their competitive positions with respect to the international division of labor. The aggregate effect is not necessarily beneficial. The transformation of the conditions of concrete labor within an urban region will, if replicated elsewhere, alter the meaning of abstract labor on the world market and so change the context in which different kinds of concrete labor are possible. Heightened competition between urban regions, like heightened competition between firms, does not necessarily lead capitalism back to some comfortable equilibrium but can spark movements that push the system farther away from it. Nevertheless, those urban regions that achieve a superior competitive position survive, at least in the short run, better than those that do not. There are, however, different paths to that end, the most important distinction being between raising the rate of exploitation of labor power (absolute surplus value) or seeking out superior technologies and organization (relative surplus value). I consider each in turn.

A shift to superior technology and organization helps particular industries within an urban region survive in the face of sharpening competition. But such a shift can just as easily eliminate jobs as create them. Growth of output and investment and decline in jobs is a familiar enough pattern (Massey and Meegan 1982). The search for superior organization can sometimes dictate radical changes in the scale of enterprise (thereby affecting the ability of firms to insert themselves into the matrix of urban possibilities, if only because of the different land needs). But it also carries over to considerations of the cost and efficiency of physical and social infrastructures. The ruling-class alliance within the urban region then has to pay much closer attention to the fine details of urban organization of cities as workshops for the production of relative surplus value. There are a number of ways it can go about that. Improved physical infrastructures and close attention to the productive forces embedded in the land (water, sewage, and so forth) improve the capacity to generate relative surplus value. But then so too do those investments in social infrastructures – education, science, and technology – that improve the urban climate as a center of innovation. Or costs to industry may be artificially reduced by subsidies. But that means redistributions of the social wage (absolute surplus value).

Sharpening interurban competition (of which there are abundant signs) poses problems. Continuous leapfrogging of technologies and organizational forms (including those provided through public investment) promotes ever fiercer competition to capture investment and jobs from highly mobile corporate capital. This has destabilizing effects and tends to accelerate the devaluation of assets and infrastructures associated with older technological mixes. Besides, accelerating technological change at the expense of growth (of output or employment) undermines the whole logic of accumulation and leads straight into the mire of global crises. Preoccupation with creating a "favorable business climate," as well as corporate handouts and other forms of subsidy to industry, can also spark popular resistance, particularly if it affects, as it usually does, the social wage. Urban politics is then in danger of reverting to class struggle rather than to more fragmented squabbles over distribution.

There are a number of checks to such immediate transitions. To begin with, the control of technology lies more within the corporation than within the innovative proclivities of the urban mix (though product innovation still retains some of its older urban base). Technology transfers between urban regions are, therefore, broadly a matter of corporate decisions. In this respect the social dominates over the spatial aspect of the division of labor. That sort of restraint, however, does not apply to infrastructure provision. Here we find the state acting as an entrepreneur (Goodman 1979) offering bait to corporate capital. And the latter is sensitive to the qualities and quantities of labor power and social infrastructures as well as to the physical resources developed within an urban region.

Raising the rate of exploitation of labor power forms another path to survival in the face of international competition in production. The classical Marxian account depicts this as a concerted attack upon labor's standard of living and an attempt to lower real wages through increased unemployment, job insecurity, the diminution of the social wage (particularly welfare provision), and the mobilization of a cheap industrial reserve army (immigrants, women, minorities, and so on). It also means an attack upon working-class institutions (such as trade unions) and the utility of skills and qualifications in employment. But this means an attack upon what may well be an important constituency of an urban-based class alliance. While we can see many an urban region moving down such a path – and in some cases urban administration has become the cutting edge for disciplining labor by wage cuts and rollbacks – there are other options that are less confrontational. The rate of exploitation is always relative, after all, to the qualities of labor power. The unique package of qualities that each urban labor market can offer, supported by selected infrastructures, can be alluring bait for mobile corporate capital. Interurban competition over quantities, qualities, and costs of labor power is, therefore, rather more nuanced than the simple version of the Marxian model would suggest. And the nuances permit a ruling-class alliance much greater flexibility to divide and rule a work force. Besides, the mobility of labor power between urban regions provides further checks to the repressive tactics through which absolute surplus value might otherwise be gained. Nevertheless, interurban competition on the labor market has a disciplining effect upon the labor force in times of faltering accumulation. The threat of job loss and of corporate flight and disinvestment, the clear need to exercise budgetary restraint in a competitive environment, point to a changing thrust of urban politics away from equity and social justice and toward efficiency, innovation, and rising real rates of exploitation (see Chap. 7).

Competition within the Spatial Division of Consumption

Urban regions can, as a second option, seek individually to improve their competitive position with respect to the spatial division of consumption. There is more to this than the redistributions achieved through tourism, important and extensive though these may be. For more than a generation, demand-side urbanization had focused heavily on life-style, the construction of community, and the organization of social space in terms of the signs and symbols of prestige, status, and power. It had also produced an ever-broadening basis for participation in such consumerism. While recession, unemployment, and the high cost of credit rendered that participation moot for important layers of the population, the game continued for the rest. Competition for their consumer dollars became more frenetic, while they, in turn, had the opportunity to become much more discriminating. The mass consumption of the 1960s was transformed into the less mass-based but more discriminating consumption of the 1970s and 1980s. Interurban competition for that consumption dollar can be fierce and costly. Investments that make for a "good living environment" and that enhance the so-called qualities of life do not come cheap. Investments seeking to establish new patterns of the spatial division of consumption are notoriously risky. Nevertheless, urban regions that successfully undertake them stand to appropriate surpluses from the circulation of revenues. And strong coalitions can be forged behind such strategies. Landlords and property owners, developers and financiers, and urban governments desperate to enhance their tax base can be joined by workers equally desperate for jobs in promoting new amusement options (of which Disney World is but a prototype), new consumer playgrounds (like Baltimore's Inner Harbor or London's docklands scheme), sports stadia and convention centers, marinas and hotels, exotic eating places and cultural facilities, and the like. The construction of totally new living environments (gentrification, retirement communities, integrated "villages in the city") fits into such a program.

But much more than physical investment is involved. The city has to appear as innovative, exciting, and creative in the realms of life-style, high culture, and fashion. Investment in support of cultural activities as well as in a wide range of urban services connects to this drive to capture surpluses from the circulation of revenues. The risks are considerable, but the pay-offs are correspondingly high. Fierce competition in this arena leads toward geopolitical struggles in the realm of cultural imperialism. The survival of cities like New York, Los Angeles, London, Paris, and Rome depends in large degree on their relative positions within this international competition for cultural hegemony and for a cut from the global circulation of revenues.

Interurban competition with respect to the spatial division of consumption has important effects. It highlights the contrast between cities as workshops for production and technological innovation and cities as centers for conspicuous consumption and cultural innovation. Serious conflicts can arise between the infrastructures necessary for these quite different functions. It also has profound implications for employment structures, emphasizing so-called services rather than blue-collar skills. And it calls for the formation of a particular kind of urban-based class alliance in which public-private co-operation in support of conspicuous consumption and cultural innovation has to play a vital role. Out of that comes a tendency, exacerbated by interurban competition, for the public subsidy of consumption by the rich at the expense of local support of the social wage of the poor. The polarizing effects of that are hard to keep in check. The argument that the only way to preserve jobs for an increasingly impoverished under class is to create consumer palaces for the rich with public subsidy has at some point to wear thin. So, too, does the ideology of a post-industrial city as the solution for capitalism's contradictions. That ideology has, however another base aside from the justification for pursuing urban survival through spatial competition for consumption. To this broader issue we now turn.

Competition for Command Functions

Urban areas can, as a third possibility, compete for those key control and command functions in high finance and government that tend, by their very nature, to be highly centralized while embodying immense power over all manner of activities and spaces. Cities can compete to become centers of finance capital, of information gathering and control, of government decision-making. Competition of this sort calls for a certain strategy of infrastructural provision. Efficiency and centrality within a worldwide network of transport and communications is vital, and that means heavy public investments in airports, rapid transit, communications systems, and the like. The provision of adequate office space and linkages depends upon a public-private coalition of property developers, financiers, and public interests capable of responding to and anticipating needs. Assembling a wide range of supporting services, particularly those that gather and process information rapidly, calls for other kinds of investments, while the specific skills requirements of such activities put a premium on urban centers with certain kinds of educational provision (business and law schools, computer training facilities, and so forth).

Competition in this realm is expensive and peculiarly tough because this is an arena characterized by monopoly power that is hard to break. The agglomeration of powerful functions in a city like New York naturally attracts other powerful functions to it. Yet, to be maximally effective command and control functions have to be hierarchically organized across space, thus imparting a powerful impulse toward hierarchical organization of the urban system as a whole (Cohen 1981). Shifts in relative spatial structures (particularly those wrought by new systems of communication) create abundant opportunities for shifts in the shape and form of the hierarchy, while new regional centers can emerge with shifts in the social and spatial divisions of labor and consumption. Indeed, command and control functions can be the cutting edge of regional readjustments and differential urban growth. And powerful advantages attach thereto. The very existence of monopolistic power permits the appropriation of surpluses produced elsewhere. And at times of economic difficulty, as Marx once observed, the financiers always tend to enrich themselves at the expense of the industrial interest simply because control over money and credit yields short-term control over the lifeblood of capitalism at a time of crisis. It is, therefore, no accident that interurban competition in the troubled years of the 1970s and 1980s focused heavily on the search to procure command and control functions at a time when there was rapid growth in such functions and multiple forces making for geographical readjustments (Friedmann and Wolff 1982).

The overall effect of such competition is to subsidize the location of command and control functions in the hope that the monopoly powers that reside therein will permit the subsidy to be recaptured through the appropriation of surplus value. That this does not necessarily help stabilize the capitalist system as a whole should be fairly self-evident. But it certainly offers a path toward individual urban survival in a world of heightened interurban competition. The effect, however, is to make it appear as if the city of the future is going to be a city of pure command and control functions, an informational city, a post-industrial city in which services lie at the heart of the urban economy.

Competition for Redistribution

Fourth, in an intricately organized society such as ours, there are many channels for direct redistribution of economic power with respect to which urban regions can and do compete. The private systems of redistribution – through organizations like the church, trade unions, professional associations, charitable organizations, and the like – are by no means negligible. Most overt interurban competition is targeted, however, on redistributions to be had from higher-order levels of government. Such expenditures grew rapidly during the Keynesian era and are still massive, though very much under attack to the degree that they were viewed by the bourgeoisie as the main culprit in inflationary deficit financing. The channels for such redistributions are, however, numerous, varied, and often hidden in obscure provisions in the tax code or in some curious executive order. The amounts that flow through these channels depend upon politics, the economy, and executive judgments. A shift in flows from one channel to another can devastate the economy of one urban region while enhancing that of another. For example, the switch from policies designed to support the social wage in the United States to deficit-financed defense expenditures after 1980 (a kind of defense-side Keynesianism) brought economic prosperity to many urban regions caught up in the defense industry. Those urban regions – located in a great arc sweeping from Connecticut and Long Island through North Carolina, Texas, and California to the state of Washington – were by no means antagonistic to the continuation of that kind of political mix.

Redistributions depend in part upon the sophistication of ruling-class alliances in procuring funds to which they might have some claim (grants for highways, sewers, education, mass transit, and so forth). But they also depend upon raw geopolitical power in relation to higher-order politics (the importance, say, of delivering the urban vote) and the threat of social unrest and political-economic disruption. The tactics of interurban competition are as varied as the modes of redistribution. The political attack on redistributive politics during the 1970s and 1980s should not be taken to mean, however, that this is no longer a viable strategy for urban survival. The city still preserves vast redistributive privileges and functions, but the mode of competition has changed quite radically since the breakdown of the Keynesian compromise.

The four options we have considered are not mutually exclusive. Happy the urban region that competes so well for the spatial division of consumption that it draws command and control functions whose high-paid personnel help capture tax redistributions for defense industries. Better still if there also exists a mix of highly skilled technocrats and a mass of recent immigrants willing to work at very low wages not only in services but also in basic production for an extensive local consumer market that forms the basis for a booming export trade. Los Angeles, for example, scored positively on all four options during the difficult years after 1973. Cities like Baltimore, Lille, and Liverpool, in contrast, scored low on most or all of them with the most dismal of results.

The coercive laws of interurban competition for the production, control, and realization of surplus value are compelling major shifts in the paths of the urbanization of capital. The forces brought to bear on urbanization are changing, but then so too is the meaning of the urban process for all aspects of economic, political, and social life. At such times of brutal and often seemingly incoherent transition, it is hard to assess that meaning, decode its complex messages, or even grasp intellectually and empirically how the variegated forces are meshing and with what consequences.

We have seen the grim headlines of capital flight, job loss, and corporate disinvestment in production against a background of rapid technological change, stuttering accumulation, a new international division of labor, a shaky international financial system, and crumbling worker power to prevent unemployment, wage cuts, and rollbacks of fringe benefits. The same headlines can be seen under the most diverse political circumstances: the United States, France, Britain, Sweden, Spain, Canada – the list is endless. Dissections of deindustrialization and programs for reindustrialization abound, as do speculations about the prospects for survival on the basis of so-called services and command functions.

The surface appearance of crisis, and therefore the focus of politicial and social concern, shifted dramatically between 1970 and 1980. Underconsumption no longer appeared as the central contradiction of capitalism, but stagflation did. The solutions to that looked quite different from those embedded in the broadly Keynesian response to the Great Depression. But behind the glamour of the high-tech industries, which are supposed to cure the problems of sagging productivity at the same time as they spawn a whole new round of product innovations, lies a reality of deep de-skilling and the routinization of boring and low-paid labor, much of it that of women. That reality was paralleled by many a journalistic exposé of the resurgence of sweatshop labor in New York, Los Angeles, London, and Paris – a different kind of solution that rested on a return to conditions of working (unregulated and tolerated) that many thought had long ago been abolished from a supposedly civilized and civilizing capitalist world. New systems of outwork, subcontracting, and home work (a wonderful way to save on direct fixed capital investment and capitalize on women's captive labor) entered upon the scene, facilitated by sophisticated systems of communication and external control. Centralization of command functions could be matchd by highly decentralized, even individualized production systems that make communication between workers difficult and so check collective consciousness and action. Behind the illusions of the post-industrial city lie the realities of a newly industrializing city. Hong Kong and Singapore are prototypes being forced back into the advanced capitalist world through interurban compe-tititon within the spatial division of labor.

We have also witnessed the headlines of glimmering hope in even the most dismal of urban regions for an urban renaissance pinned together out of some mix of office development, entertainment centers, shopping malls, and investment in new living environments and gentrification of the old. Some cities present such a glamorous and dynamic face to the world that it is hard to credit some of the realities that lie within. In New York City, that amazing center of immensely centralized economic power, cultural imperialism, conspicuous consumption, and dramatic gentrification (Soho, the Upper West Side, even into Harlem), one in four households now ekes out a living on incomes below the poverty level, and one out of every two children is raised under similar conditions. The supply of affordable housing for an increasingly impoverished population in Baltimore is worse now than in the 1960s. Yet Baltimore is touted as a national, even an international model of urban renaissance built upon tourism and increasingly conspicuous consumption. Curiously, the headlines of housing deprivation, hunger, lack of access to medical care and education, injustices of distribution, and discrimination based on race, gender, and place have lost the primacy they had in the supposed urban crisis of the 1960s, even though the conditions now are worse than they were then. If the question of distribution is placed upon the political agenda at all, it is in terms of restructuring material incentives to the enterprising and diminishing the power of labor in order to confront a sagging ability to produce rather than realize surplus value. From that follows the savage attack in some advanced capitalist countries (principally Britain and the United States) upon the welfare state. But interurban competition, by concentrating on subsidies to corporations and upper-income consumption, feeds that process of polarization at the local level in powerful ways. Capitalist urbanization thereby drops its seemingly human mask. We turn back to a style of capitalist urbanization that the Keynesian social planners struggled so gamely to reverse after World War II. The rich now grow richer and the poor grow poorer, not necessarily because anyone wills it that way (though there are plainly those in power who do), but because it is the natural outcome of the coercive laws of competition. And within the many dimensions of the heightened competition, interurban competition has a powerful role to play.

VI. THE URBANIZATION OF CAPITAL

Henri Lefebvre has long argued, somewhat in the wilderness it must be admitted, that the urban process has more importance in the dynamics of capitalism that most analysts are ever prepared to contemplate. The studies I have undertaken these last few years on the history and theory of the urbanization of capital bear witness to the cogency of Lefebvre's message. They do so on a number of counts.

Urbanization has always been about the mobilization, production, appropriation, and absorption of economic surpluses. To the degree that capitalism is but a special version of that, we can reasonably argue that the urban process has more universal meaning than the specific analysis of any particular mode of production. This is, of course, the track that much comparative urban study has taken. But urbanization is used under capitalism in very specific ways. The surpluses sought, set in motion and absorbed are surpluses of the product of labor (appropriated as capital and usually expressed as concentrated money power) and surpluses of the capacity to labor (expressed as labor power in commodity form). The class character of capitalism dictates a certain manner of appropriation and a split of the surplus into the antagonistic and sometimes mutually irreconcilable forms of capital and labor. When the antagonism cannot be accommodated, capitalism has to add powers of devaluation and destruction of both capital and labor surpluses to its lexicon of possibilities. Powerfully creative in many ways – particularly with respect to technology, organization, and the ability to transform material nature into social wealth – the bourgeoisie also has to face up to the uncomfortable fact that it is, as Berman (1982, 100) puts it, "the most destructive ruling class in world history." It is the master of creative destruction. The class character of capitalism radically modifies the manner and meaning of the mobilization, production, appropriation, and absorption of economic surpluses. The meaning of urbanization is likewise radically redefined.

It is always tempting when faced with categories of this sort to turn them into "historical stages" of capitalist development. I have taken such a path in this chapter to some degree by pointing to the mobilization of surpluses in the mercantile city, the production of surpluses in the industrial city, and the absorption of surpluses in the Keynesian city as pegs on which to hang an abbreviated account of the history of capitalist urbanization. In truth, matters are somewhat more complicated and nuanced. Though the emphasis may vary, appropriation, mobilization, production and absorption are ever separate moments in an integrated process. How they hang together in space and time is what counts. A reconstruction of the temporal and spatial dynamics of capital circulation under the specific class relations of capitalism indicates the points of integration for a capitalist mode of production. But as we saw in the case of urbanization in the post-Keynesian era of transition, all kinds of mixes of strategies are possible, given the particular form of urban organization and economy in the context of its space relations.

While urbanization might reasonably be presented as an expression of all that, we have also to recognize that it is through urbanization that the surpluses are mobilized, produced, absorbed, and appropriated and that it is through urban decay and social degradation that the surpluses are devalued and destroyed. And like any means, urbanization has ways of determining ends and outcomes, of defining possibilities and constraints, and of modifying the prospects for capitalist development as well as for the transition to socialism. Capitalism has to urbanize in order to reproduce itself. But the urbanization of capital creates contradictions. The social and physical landscape of an urbanized capitalism is far more, therefore, than a mute testimony to the transforming powers of capitalist growth and technological change. Capitalist urbanization has its own distinctive logic and its own distinctive forms of contradiction.

The grounds for that can be established by a different path. There are, I submit, immense gains to be had from looking closely at the rich complexity and intricately woven textures of urban life as the crucible for much that is fundamental to human experience, consciousness formation, and political action. I take up such matters at much greater length in *Consciousness and the Urban Experience,* but I cannot let them pass without some commentary here. The study of urban life illuminates people in multiple roles – workers, bosses, homemakers, consumers, community residents, political activists, borrowers, lenders, and so forth. The roles do not necessarily harmonize. Individuals internalize all kinds of stresses and strains, and external signs of individual and collective conflict abound. But urbanization means a certain mode of human organization in space and time that can somehow embrace all of these conflicting forces, not necessarily so as to harmonize them, but to channel them into so many possibilities of both creative and destructive social transformation. There is plainly much more at stake here than mere class interest. Yet capitalist urbanization presupposes that the urban process can somehow be mobilized into configurations that contribute to the perpetuation of capitalism. How can that be? The short answer is quite simply that it is not necessarily so. The urban form of organization that capitalism implants does not necessarily adapt to every dictate of that mode of production any more than individual or collective consciousness boils down to simple and polarized class struggle.

Such dilemmas lurk in the various strategies for urban survival in the post-Keynesian transition. The search to produce surpluses in one place depends on the ability to realize and absorb them in another. The mobilization of surpluses via command functions presumes there is some production somewhere to command. The overall stability of capitalism depends on the coherence of such integrations. Yet urban-based class alliances (even when themselves coherently organized) do not form and strategize in relation to such global considerations of coordination. They compete to save their own asset base as best they can and to preserve their power of appropriation whatever way they can. To be sure, corporate and finance capital and, to a lesser degree, labor power are mobile across urban entities (thus rendering the urban-based class alliances permanently vulnerable). But this does not guarantee an urban evolution exactly geared to capitalism's requirements. It simply emphasizes the ever-present tension between the social and spatial divisions of production, consumption, and control.

Interurban competition is, then, one important determinant in capitalism's evolution and is fundamental, as I argued in Chapter 6, to its uneven geographical development. That competition could be viewed as potentially harmonious if Adam Smith was right that the hidden hand of the market invariably transforms individual selfishness, ambition, and short-sightedness into a global social outcome that benefits all. But Marx's devastating rebuttal of that thesis prevails here too. The more perfect the hidden hand of interurban competition, the more the inequality between capital and labor builds and the more unstable capitalism becomes. Heightened competition is a way into rather than out of capitalist crisis in the long run.

What, then, is the post-Keynesian transition a transition to? That is a question to which there is no automatic answer. The laws of motion of capitalism track the underlying contradictions that push capitalism to evolve, but they do not dictate the paths to take. Our historical geography is always ours to make. But the conditions under which we seek to make that historical geography are always highly structured and constrained. Viewed solely from the standpoint of interurban competition, for example – and I admit this is a drastic simplification that I shall not even try to justify – there is much to indicate spiraling temporal disequilibrium within a rapidly seesawing movement of uneven geographical development; sporadic place-specific devaluations coupled with even more sporadic bursts of place-specific accumulation. And there is more than a little evidence to support that. The Sun Belt cities in the United States that rode so high and secure on the energy boom after 1973 slip rapidly into depression with every drop in oil price – Houston, Dallas, and Denver, once boom towns, are now in deep trouble. High-tech centers like Silicon Valley turn rapidly sour, while New York City, which seemed on the point of total collapse in the early 1970s, suddenly adds command-type functions and even low-wage manufacturing jobs oriented to the local market. These are the kinds of rapid shifts in fortune that we would expect to see under conditions of heightened interurban competition for the mobilization, production, appropriation, and absorption of surpluses.

But are there any broader indicators? The emphasis on command and consumption in the United States puts the focus on appropriation rather than on production, and in the long run that spells acute geopolitical danger as more and more cities become centers of mercantilist endeavor in a world of shrinking profitable production possibilities. This was the kind of volatile mix that, at the nation state level, led straight into those lopsided patterns of uneven geographical development characteristic of the age of high imperialism. And that was the kind of tension that lay at the root of two world wars. Yet, the search for profitable production possibilities under conditions of heightened competition between firms, urban regions, and nations points to rapid transitions in the sociotechnical and organizational conditions of production and consumption. And that portends disruption of whatever structured coherence has been achieved within an urban economy, substantial devaluation of many of the physical and social infrastructural assets built up there, and instability within any ruling-class alliance. It also means destruction of many traditional skills within the labor force, the devaluation of labor power, and disruption of powerful cultures of social reproduction. Bringing the Third World back home is not an easy follow-up to Keynesian-style urbanization. Ironically, moving too rapidly down that path also dramatizes the crisis tendencies of capitalism as underconsumption problems once more.

What, then, of the possibilities of transition to some alternative mode of production and consumption? At a time when the struggle for survival within capitalism dominates political and economic practice and consciousness, it becomes doubly hard to think about a radical break and the construction of a socialist alternative. Yet the present insecurities and instabilities, to say nothing of the threat of massive devaluation and destruction through internal reorganization, geopolitical confrontation, and political-economic breakdown, make the question more vital than ever.

The alternative cannot, however, be constructed out of some unreal socialist blueprint. It has to be painfully wrought through a transformation of society, including its distinctive forms of urbanization, as we know it. A study of the urbanization of capital indicates the possibilities and the necessary constraints that face struggle toward that goal. The historical geography of capitalism has shaped physical and social landscapes in profound ways. These landscapes now form the humanly created resources and productive forces and mirror the social relations out of which socialist configurations will have to be carved. The uneven geographical development of capitalism can at best be slowly modified and the maintenance of existing spatial configurations – so essential to the reproduction of social life as we now know it – means the continued structuration and replication of spaces of domination and subservience, of advantage and disadvantage. How to break out of that without destroying social life is the quintessential question. The urbanization of capital imprisons us in myriad and powerful ways. Like any sculptor, we are necessarily limited by the nature of the raw material out of which we try to build new shapes and forms. And we have to recognize that the physical and social landscape of capitalism as structured through its distinctive form of urbanization contains all manner of hidden flaws, barriers, and prejudices inimical to the construction of any idealized socialism.

But capitalism is also destructive of all that, perpetually revolutionizing itself and always teetering on that knife-edge of preserving its own values and traditions and necessarily destroying them to open up fresh room for accumulation. What Henry James called "the reiterated sacrifice to pecuniary profit" makes the urbanization of capital a peculiarly open and dynamic affair. The urban is, consequently, as Lefebvre (1974) is fond of saying, "the place of the unexpected"; and out of that all manner of possibilities can flow. The problem is to understand the possibilities and create the political instrumentalities appropriate to their exploitation. The tactics of working-class struggle have to be as fluid and dynamic as capitalism itself. The shift, for example, toward a more corporatist style of urbanization in the United States in the period of the post-Keynesian transition opens a space into which movements toward municipal socialism can more readily be inserted to form the basis for broader political struggle. But for that opportunity to be seized requires a radical transition in American urban politics away from fragmented pluralism into a more class-conscious mode of politics. The barriers to that process, as I show in *Consciousness and the Urban Experience,* are profound indeed because they are deeply embedded in the structures of contemporary capitalism itself. The individualism of money, the consciousness of family and community, the chauvinism of state and local governments, compete with the experience of class relations on the job and create a cacophony of conflicting ideologies which all of us to some degree internalize.

But even presupposing that consciousness of class emerges supreme within the complex rivalries of urban social movements, there is another whole dimension to struggle that has to be confronted. It is noticeable, for example, that in those European countries in which municipal socialism has already won its laurels and where a more articulate class-based politics does indeed prevail, that the corporatist powers of the urban-based class alliance are whittled away and replaced by the powers of the nation state where the bourgeoisie can more easily retain its power. The allocation of powers between urban region, state, and multinational organs is itself an outcome of class struggle. The bourgeoisie will always seek to shift authority, powers, and functions away from the spaces it cannot control into the spaces within which its hegemony prevails. The tension between city and state that Braudel (1984) makes so much of in his description of the rise of capitalism is still with us. It deserves more careful scrutiny as part and parcel of the processes of class struggle around the survival of capitalism and the production of socialism. Capitalism has survived not only through the production of space, as Lefebvre insists, but also through superior command over space – and that truth prevails as much within urban regions as over the global space of capitalist endeavor.

The urbanization of capital is but a part of the total complex of problems that confronts us in the search for an alternative to capitalism. But it is a vital part. An understanding of how capital becomes urbanized and the consequences of that urbanization is a necessary condition for the articulation of any theory of the transition to socialism. In the final paragraph of *Social Justice and the City* I wrote these lines:

> A genuinely humanizing urbanism has yet to be brought into being. It remains for revolutionary theory to chart a path from an urbanism based in exploitation to an urbanism appropriate for the human species. And it remains for revolutionary practice to accomplish such a transformation.

That aim still stands. But I would now want to put it in a broader perspective. Any movement toward socialism that does not confront the urbanization of capital and its consequences is bound to fail. The construction of a distinctively socialist form of urbanization is as necessary to the transition to socialism as the rise of the capitalist city was to the sustenance of capitalism. Thinking through the paths to socialist urbanization is to chart the way to the socialist alternative itself. And that is what revolutionary practice has to accomplish.

References

WORKS BY MARX

The poverty of philosophy. International Publishers, New York, 1963.

The eighteenth brumaire of Louis Bonaparte. International Publishers, New York, 1963.

Wages, price and profit. Foreign Languages Press, Peking, 1965.

Capital. 3 vols. International Publishers, New York, 1967.

Critique of Hegel's philosophy of right. Ed. J. O'Malley, Cambridge University Press, London, 1970.

Theories of surplus value. 3 vols. Lawrence and Wishart, London, pts. 1 and 2, 1969; pt. 3, 1972.

Grundrisse. Ed. M. Nicolaus. Penguin Publishers, Harmondsworth, Middlesex, 1973.

The results of the immediate process of production. Appendix to *Capital,* vol. 1. Ed. E. Mandel. Penguin Publishers, Harmondsworth, Middlesex, 1976.

WORKS BY MARX AND ENGELS

Manifesto of the Communist Party. Progress Publishers, Moscow, 1952.

Selected correspondence. Progress Publishers, Moscow, 1955.

The German ideology. Ed. C. J. Arthur. International Publishers, New York, 1970.

On colonialism. International Publishers, New York, 1972.

Ireland and the Irish question. Prepared by L. I. Goldman and V. E. Kumina. International Publishers, New York, 1972.

OTHER WORKS CITED

Abramovitz, M. 1964. *Evidences of long swings in aggregate construction since the Civil War.* New York.

Albert, W. 1972. *The turnpike road system in England.* London.

Alonso, W. 1964. *Location and land use.* Cambridge, Mass.

Altvater, E. 1973. Notes on some problems of state interventionism. *Kapitalistate* 1:96–108.

Ambrose, P., and R. Colenutt. 1975. *The property machine.* Harmondsworth, Middlesex.

Barker, G., J. Penney, and W. Seccombe. 1973. *Highrise and superprofits.* Kitchener, Ontario.

Bender, T. 1975. *Toward an urban vision: Ideas and institutions in nineteenth-century America.* Lexington, Ky.

Berman, M. 1982. *All that is solid melts into air.* New York.

Berry, B., and E. Neils. 1969. Location, size, and shape of cities as influenced by environmental factors: The urban environment writ large. In *The quality of the urban environment,* ed. H. Perloff. Baltimore.

Blaut, J. 1974. The ghetto as an internal neo-colony. *Antipode* 6, no. 1:37–41.

Boyer, B. 1973. *Cities destroyed for cash.* New York.

Brace, C. L. 1889. *The dangerous classes of New York.* New York.

Braudel, F. [1979] 1984. *The perspective of the world.* Trans. S. Reynolds. New York.

Bunge, W. 1971. *Fitzgerald: The geography of a revolution.* Cambridge, Mass.

Castells, M. 1972. *La question urbaine.* Paris.

———. 1975. Collective consumption and urban contradictions in advanced capitalist societies. In *Patterns of advanced societies,* ed. L. Lindberg. New York.

———. 1983. *The city and the grassroots.* Berkeley and Los Angeles.

Castells, M., and F. Godard. 1973. *Monopolville: L'enterprise, l'état, l'urbain.* The Hague.

Chalmers, T. 1821–26. *The Christian and civil economy of large towns.* 3 vols. Glasgow.

Chatterjee, L. 1973. Real estate investment and deterioration of housing in Baltimore. Ph.D. diss., Department of Geography and Environmental Engineering, Johns Hopkins University, Baltimore.

Chinitz, B. 1958. Contrasts in agglomeration: New York and Pittsburgh. *American Economic Review* 51:279–89.

Cloward, R., and F. Piven. 1974. *The politics of turmoil.* New York.

Cohen, R. 1981. The new international division of labor, multinational corporations, and urban hierarchy. In *Urbanization and urban planning in capitalist society,* ed. M. Dear and A. Scott. London.

Coing, H. 1982. *La ville, marché de l'emploi.* Paris.

Counter-Information Services. 1973. *Anti-report on the property developers.* London.

Day, R. 1976. The theory of long waves: Kondratieff, Trotsky, Mandel. *New Left Review* 99:67–82.

Deane, P., and W. Cole. 1967. *British economic growth, 1688–1959.* London.

Dickens, C. [1854] 1961. *Hard times.* Reprint. New York.

———. [1846–48] 1970. *Dombey and son.* Reprint. Harmondsworth, Middlesex.

Downie, L. 1974. *Mortgage on America.* New York.

Engels, F. [1872] 1935. *The housing question.* International Publishers, New York.

———. [1845] 1971. *The condition of the working class in England in 1844.* 2d ed., trans. and ed. W. O. Henderson and W. H. Challoner. London.

Fine, B. 1979. On Marx's theory of agricultural rent. *Economy and Society* 8:241–78.

Firey, W. 1960. *Man, mind, and the land: A theory of resource use.* Glencoe, Ill.

Foster, J. 1974. *Class struggle in the industrial revolution.* London.

Friedmann, J., and G. Wolff. 1982. World city formation: An agenda for research. *International Journal of Urban and Regional Research* 6:309–44.

Gaffney, M. 1973. Releasing land to serve demand via fiscal disaggregation. In *Modernizing urban land use policy*, ed. M. Clawson. Washington, D.C.

Giddens, A. 1973. *The class structure of the advanced societies*. London.

———. 1981. *A contemporary critique of historical materialism*. London.

Giglioli, P., ed. 1972. *Language and social context*. Harmondsworth, Middlesex.

Girard, L. 1952. *Les politiques des travaux publics sous le Second Empire*. Paris.

Godelier, M. [1966] 1972. *Rationality and irrationality in economics*. Trans. B. Pearce. London.

Goodman, R. 1971. *After the planners*. New York.

———. 1979. *The last entrepreneurs*. Boston.

Gottlieb, M. 1976. *Long swings in urban development*. New York.

Gramsci, A. 1971. *Selections from the prison notebooks*. Trans. and ed. Q. Hoare and G. N. Smith. London.

Grigsby, W., L. Rosenberg, M. Stegman, and J. Taylor. 1971. *Housing and poverty*. Philadelphia.

Habermas, J. [1973] 1975. *Legitimation crisis*. Trans. T. McCarthy. Boston.

Hall, P., H. Gracey, R. Crewett, and R. Thomas. 1973. *The containment of urban England*. 2 vols. London.

Hareven, T. 1982. *Family time and industrial time*. London.

Harvey, D. 1973. *Social justice and the city*. London.

———. 1975. The political economy of urbanization in the advanced capitalist societies – the case of the United States. In *The social economy of cities*, ed. G. Gappert and H. Rose. Annual Review of Urban Affairs no. 9. Beverly Hills.

———. 1981. The spatial fix: Hegel, von Thunen, and Marx. *Antipode* 13, no. 3:1–12.

———. 1982. *The limits to capital*. Oxford.

———. 1985. The geopolitics of capitalism. In *Social relations and spatial structures*, ed. D. Gregory and J. Urry. London.

Harvey, D., and L. Chatterjee. 1974. Absolute rent and the structuring of space by financial institutions. *Antipode* 6, no. 1:22–36.

Hawley, A., and O. Duncan. 1957. Social area analysis. *Land Economics* 33:340–51.

Hegel, G. W. [1821] 1952. *Philosophy of right*. Trans. T. M. Knox. Oxford.

Herman, E. 1973. Do bankers control corporations? *Monthly Review* 25, no. 2:12–29.

Holton, R. 1984. Cities and the transition to capitalism and socialism. *International Journal of Urban and Regional Research* 8:13–37.

Houdeville, L. 1969. *Pour une civilisation de l'habitat*. Paris.

Isard, W. 1942. A neglected cycle: The transport building cycle. *Review of Economics and Statistics* 24:149–58.

Jacobs, J. 1969. *The economy of cities*. New York.

———. 1984. *Cities and the wealth of nations*. New York.

Katznelson, I. 1981. *City trenches: Urban politics and the patterning of class in the United States*. New York.

Keiper, J. S., E. Kurnow, C. Clark, and H. Segal. 1961. *Theory and measurement of ground rent*. Philadelphia.

Kuznets, S. 1961. *Capital in the American economy: Its formation and financing*. Princeton.

Lefebvre, H. 1970. *La révolution urbaine*. Paris.

———. 1972. *Le droit à la ville*. Paris.

———. 1974. *La production de l'espace*. Paris.

Lewis, J. P. 1965. *Building cycles and Britain's growth*. London.

Lukács, G. [1923] 1968. *History and class consciousness*. Trans. R. Livingstone. London.

McPherson, C. B. 1962. *The political theory of possessive individualism*. London.

Malthus, T. [1836] 1951. *The principles of political economy*. Reprint. New York.

Mandel, E. [1972] 1975. *Late capitalism*. Trans. J. de Brés. London.

Marcuse, H. 1968. *Negations: Essays in critical theory*. Boston.

Marriott, O. 1967. *The property boom*. London.

Marx, K., and V. I. Lenin. 1968. *The civil war in France: The Paris Commune*. International Publishers, New York.

Massey, D., and A. Catelano. 1978. *Capital and land: Land ownership by capital in Great Britain*. London.

Massey, D., and R. Meegan. 1982. *The anatomy of job loss*. London.

Meier, R. 1962. *A communications theory of urban growth*. Cambridge, Mass.

Merrington, J. 1975. Town and country in the transition to capitalism. *New Left Review* 93:73–92.

Miliband, R. 1968. *The state in capitalist society*. London.

Milner-Holland Report. 1965. *Report of the Committee on Housing in Greater London*. London.

Mollenkopf, J. 1983. *The contested city*. Princeton.

Molotch, H. 1976. The city as a growth machine: Toward a political economy of place. *American Journal of Sociology* 82:309–32.

Mumford, L. 1961. *The city in history*. New York.

National Resources Committee (of the United States). 1937. *Our cities: Their role in the national economy*. Washington, D.C.

Neutze, M. 1968. *The suburban apartment boom*. Baltimore.

Newson, E., and J. Newson. 1970. *Four years old in an urban community*. Harmondsworth, Middlesex.

O'Connor, J. 1973. *The fiscal crisis of the state*. New York.

Ollman, B. 1971. *Alienation: Marx's conception of man in capitalist society*. London.

———. 1973. Marxism and political science: A prolegomenon to a debate on Marx's method. *Politics and Society* 3:491–510.

Pahl, R. [1970] 1975. *Whose city?* Reprint. Harmondsworth, Middlesex.

Pickvance, C., ed. 1976. *Urban sociology: Critical essays*. London.

Piven, F., and R. Cloward. 1971. *Regulating the poor*. New York.

Pollard, S. 1965. *The genesis of modern management*. Cambridge, Mass.

Postan, M. 1935. Recent trends in the accumulation of capital. *Economic History Review* 6:1–12.

Postel-Vinay, G. 1974. *La rente foncière dans le capitalisme agricole*. Paris.

Poulantzas, N. [1968] 1973. *Political power and social classes*. Trans. and ed. T. O'Hagen. London.

Preteceille, E. 1975. *Equipements collectifs, structures urbaines, et consommation sociale*. Paris.

Rex, J., and R. Moore [1967] 1975. *Race, community, and conflict*. Reprint, 2d ed. London.

Rey, P-P. 1973. *Les alliances de classes*. Paris.

Robson, B. 1969. *Urban analysis: A study in city structure*. London.

Rougerie, J. 1968. Remarques sur l'histoire des salaires à Paris au dix-neuvième siècle. *Le Mouvement Sociale* 63:71–108.

Saunders, P. 1981. *Social theory and the urban question*. New York.

Smith, N. 1984. *Uneven development: Nature, capital and the production of space*. Oxford.

Spoehr, A. 1956. Cultural differences in the interpretation of natural resources. In *Man's role in changing the face of the earth*, ed. W. Thomas. Chicago.

Sternlieb, G. 1966. *The tenement landlord*. New Brunswick, N.J.

Storper, M., and R. Walker. 1983. The theory of labor and the theory of location. *International Journal of Urban and Regional Research* 7:1–43.

———. 1984. The spatial division of labor: Labor and the location of industries. In *Sunbelt/Snowbelt*, ed. L. Sawers and W. Tabb. New York.

Tarr, J. 1973. From city to suburb: The "moral" influence of transportation technology. In *American Urban History*, ed. A. Callow. New York.

Thomas, B. 1972. *Migration and economic growth: A study of Great Britain and the Atlantic economy*. 2d ed., rev. London.

Timms, D. 1971. *The urban mosaic: Towards a theory of residential segregation*. London.

United States House of Representatives, Committee on Banking (Staff Report). 1968. *Trust banking in the United States*. Washington, D.C.

United States House of Representatives, Judiciary Committee (Staff Report). 1971. *Report on conglomerates*. Washington, D.C.

Walker, R. A. 1976. The suburban solution. Ph.D. diss., Department of Geography and Environmental Engineering, Johns Hopkins University, Baltimore.

———. 1981. A theory of suburbanization. In *Urbanization and planning in capitalist society*, ed. M. Dear and A. Scott. New York.

Ward, J. R. 1974. *The finance of canal building in the eighteenth century*. London.

Wicksteed, P. 1894. *The coordination of the laws of distribution*. London.

Williams, R. 1973. *The country and the city*. London.

Wirth, L. 1964. *On cities and social life*. Ed. A. J. Reiss, Jr. Chicago.

Zukin, S. *Loft living*. Baltimore, 1982.

Index